Controlling tropical deforestation

Alan Grainger

Earthscan Publications Ltd, London

First published 1993 by
Earthscan Publications Ltd
120 Pentonville Road, London N1 9JN

British Library Cataloguing-in-Publication Data

A catalogue record for this book is available from the British Library

ISBN 1 85383 142 5

Typeset by Books Unlimited (Nottm) – Sutton in Ashfield, NG17 1AL
Printed by Biddles Ltd, Guildford and Kings Lynn.

Earthscan Publications Ltd is an editorially independent subsidiary of Kogan
Page Limited and publishes in association with the International Institute for
Environment and Development and the World Wide Fund for Nature.

Contents

List of figures

9

List of tables

10

Acronyms

AVHRR	Advanced very high resolution radiometer
CDIAC	Carbon Dioxide Information Analysis Centre, USA
CVRD	Companhia Vale do Rio Doce
dbh	Diameter at breast height
EC	European Community
ECU	Environmental Change Unit, University of Oxford
FAO	United Nations Food and Agriculture Organization
GATT	General Agreement on Tariffs and Trade
GCM	Global circulation model
GEF	Global Environmental Facility
GEMS	Global Environment Monitoring System
GNP	Gross national product
gt	Gigatonne
GTZ	German Agency for Technical Cooperation
GW	Gigawatt
IBAMA	Ministry of Environment and Natural Resources, Brazil
IBDF	Ministry of Forestry, Brazil (former)
ICRAF	International Council for Research in Agroforestry
IIASA	International Institute for Applied Systems Analysis
IIED	International Institute for Environment and Development
IITA	International Institute for Tropical Agriculture
INBIO	National Biodiversity Institute, Costa Rica
INPE	Brazilian National Space Institute

IPCC	Intergovernmental Panel for Climate Change
ITTA	International Tropical Timber Agreement
ITTO	International Tropical Timber Organization
IUCN	International Union for the Conservation of Nature/World Conservation Union
MSS	Multi-spectral scanner
MW	Megawatt
NAS	US National Academy of Sciences
NASA	US National Aeronautic and Space Administration
NGO	Non-governmental organization
NOAA	US National Oceanographic and Atmospheric Administration
NRC	US National Research Council
OPEC	Organization of Petroleum Exporting Countries
PAD	Directed settlement project (Brazil)
PIC	Integrated colonization project (Brazil)
PIN	National Integration Programme (Brazil)
POLONOROESTE	Northwest Brazil Integrated Development Scheme
ppm	Parts per million
RFF	Resources for the Future
rwe	Roundwood equivalent volume
SUDAM	Superintendency for the Development of the Amazon (Brazil)
TFAP	Tropical Forestry Action Plan/Programme
TROPFORM	Tropical Forest Simulation Model
TM	Thematic mapper
UN	United Nations Organization
UNCED	UN Conference on Environment and Development
UNCTAD	UN Conference on Trade and Development
UNDP	UN Development Programme
UNEP	UN Environment Programme
UNESCO	United Nations Educational, Scientific and Cultural Organization
USAID	US Agency for International Development
USDA	US Department of Agriculture
WCS	World Conservation Strategy
WEAP	World Ecological Areas Programme

WMO	World Meteorological Organization
WRI	World Resources Institute
WWF	Worldwide Fund for Nature (still known as the World Wildlife Fund in the USA)

Preface

The deforestation of tropical rain forests has been a source of world-wide concern for more than twenty years. This book sets out to give a concise but comprehensive introduction to the causes, scale and possible consequences of the problem and to suggest how it might be tackled. It is intended to satisfy the needs of general readers and students on introductory courses who want to gain a quick overview of tropical deforestation, and to provide some new ideas and inspiration to politicians, planners, agronomists, foresters, conservationists and development workers keen to grapple with the major policy and technical challenges that tropical deforestation presents to them. It is important to stress that deforestation is also happening elsewhere in the tropics, particularly in dry areas, but this merits separate treatment.

This is a fascinating and complex subject, with many links to a wide variety of other topics, so in order to keep the book to a reasonable length I have had to be extremely ruthless in selecting which material to include. To those readers who, on reaching the last page, find that I have failed to mention this or that topic, I can only point out that there are many other interesting matters not referred to either, but if they all had been included the book would have been three or four times as long. Fellow researchers, advanced students and other readers requiring a more detailed, academic treatment are referred to another, and much more extensive forthcoming book (Grainger, 1993).

I am grateful to the many people and organizations who have supported my research over the years, especially the late Richard St Barbe Baker, founder of The Men of the Trees, who helped to pioneer the notion that forests matter on a global scale and inspired me to work in this field; Edward Goldsmith, who commissioned my 1980 report in *The Ecologist*; and, most of all, the late Lawrence and Cherry Hills, and the International Tree Crops Institute, who so generously

15

funded my research at the Oxford Forestry Institute between 1981 and 1986. Philip Stewart and Duncan Poore supervised that work; the International Federation of Institutes of Advanced Study funded the associated field studies in Borneo through its programme headed by Gordon Hallsworth; and those studies were made possible by the practical support of the late Michael Ross, the German Agency for Technical Cooperation and the kind cooperation of many officials and scientists in Indonesia and Malaysia. My thanks also go to Resources for the Future, which awarded me a Gilbert F. White Fellowship, and to Roger Sedjo for encouraging me to apply for it and being a stimulating colleague during my tenure. In recent years I have greatly benefited from the support and inspiration provided by my colleagues and students at the universities of Salford, Stirling, and now Leeds. Finally, I wish to thank Lois Wright, Alison Manson and Gustav Dobrzynski for their help in preparing the illustrations, and an anonymous reviewer, Michael Eden and Duncan Poore for their valuable comments on the manuscript. Any errors which remain are entirely my responsibility.

The tropical rain forests are one of the great natural wonders of the world, and a compelling attraction for people from many countries who, even if they have never seen or visited them, know that something precious is being lost as a result of tropical deforestation and that the repercussions will affect them. Surely something must be done? Indeed, but tropical deforestation is caused by people just like us, behaving as naturally as we do every day when we try to earn our living as best we can. We cannot deny this reality. If we are to save the forests we have to understand why deforestation is happening and go on from there to propose solutions that are realistic to people in the humid tropics and enhance their lives in a meaningful way. The solutions suggested here may not be as quick and easy as some would like but I hope they will be found feasible by those who control the fate of the tropical rain forests – the people of the humid tropics.

Introduction

The tropical rain forests are falling at human hands. The world's most valuable ecosystem is under threat because of our need for farmland, timber, minerals and other resources. These forests, which spread over more than sixty countries, account for almost two fifths of all closed forests in the world and contain half of all plant and animal species. Yet deforestation is happening on such a scale that if it were to continue at present rates the forests could disappear within the next one to two hundred years.

Why should the fate of these faraway forests concern us? If the forests go, then so will many of the species they support. The biological diversity of the whole planet will be severely reduced, threatening the continued cultivation of many important crops that came originally from wild plants in the tropical rain forests, and curtailing our future options to exploit other plants for medicines and other products. The cultures of indigenous peoples who live inside the forests will be destroyed. There will be widespread soil degradation, changes in river regimes and increased sedimentation of reservoirs. Future supplies of tropical hardwood will be put at risk. Finally, deforestation will contribute to impending global climate change through the greenhouse effect. For all these reasons it is vital to control deforestation so that a large area of tropical rain forest remains. But how can this be done?

The causes of tropical deforestation

The first thing to do is to identify the causes of the problem. There has been widespread concern about tropical deforestation for over twenty years yet so far little progress has been made in controlling it. Perhaps this is because we have all been too eager to propose solutions without first identifying the causes – which are by no means simple or easy to influence. The forests are being cleared because people want to use the land to grow food, extract minerals or build hydroelectric schemes instead. They are also being logged to supply demand for timber at home and abroad.

Demand for food, timber and other commodities results in turn from a whole host of underlying social, economic and political causes, including population growth, economic development, poverty and inequality, the desire of governments to 'civilize' remote wilderness areas, and foreign demand for food, cash crops, timber and minerals. Some deforestation is planned by governments as a deliberate act of policy but much occurs spontaneously and is outside their control. If we are to control deforestation we must recognize the crucial importance of these underlying causes and adopt an integrated approach that embraces them and the wide range of land uses that replace forest – and not just selected ones like shifting cultivation.

Land use change and economic development

In the eyes of many people living in tropical countries both deforestation and logging seem entirely justified. They are viewed as inevitable consequences of the changes in land use patterns that take place as a largely forested country develops. We might question the validity or desirability of economic development, but that is the subject for another book. Because of its link with development deforestation is not confined to the humid tropics but is also occurring throughout the developing world: in dry tropical regions such as the Sahel in West Africa where it is contributing to desertification (Grainger, 1990a), and in montane areas such as the Himalayas. However, this book is focusing on that part of tropical deforestation taking place in the tropical moist forests.

Deforestation is not unique to the tropics: historically, people all over the world have cleared forest to make productive use of the land beneath it. In some countries, such as the UK, virtually all forest was cleared. In others, such as the USA, deforestation was also extensive and rapid but was eventually brought under control before it progressed as far. There are many parallels between historical patterns of deforestation in the temperate world and current trends in the humid tropics. The main reasons why extensive clearance of tropical rain forests has been delayed so long are that environmental conditions made settlement unattractive and rapid economic development is a more recent phenomenon in that part of the world.

If some deforestation is indeed inevitable in the humid tropics it would seem futile to attempt to halt it completely within the space of a few years. A more realistic target would be to exert ever tighter control of deforestation by tackling the causes rather than the symptoms of the problem, so that future rates of forest loss decline and are kept at manageable levels. We can do this by ensuring that land use becomes as sustainable and productive as possible and improving people's lives so the pressures to clear forest are reduced. By tackling the problem in this way the prospects for long-term success will be much brighter.

The role of forestry

If clearing forest for agriculture is justifiable in the eyes of local people, then so too is retaining it for long-term timber production. Throughout history all countries in the world have struggled to find the right balance between forests and farming. More recently, and particularly in tropical areas, another option became available: conserving forest free from any use in national parks and other protected areas. But the main reason why many temperate countries still retain large areas of forest today was that governments and individuals realized that they had value as renewable sources of timber. Timber production may not be as economically attractive as other land uses in the short term but timber is still an essential commodity and forests are needed to produce it. Logging in tropical rain forests, like deforestation, has accelerated since the Second World War, exploiting the vast quantities of timber that they contain to supply the needs of people at home and overseas. If managed properly and protected from deforestation these forests could in future supply an increasing share of world timber production as demand for wood products in the developed world gets ever closer to the sustainable supply limits of temperate forests and demand accelerates in the developing world.

But how can this be reconciled with the need to conserve the full natural wealth of tropical rain forest ecosystems? One advantage is that because only a small number of the many thousands of tree species in the tropical rain forests are commercially marketable at the moment, logging only removes a few trees per hectare rather than clearing the whole forest. This practice, called selective logging, is found almost universally in tropical rain forests, in contrast to the clear-felling common in temperate forests. Some deforestation is needed to build logging roads, camps, etc. but the forest still remains after the loggers leave, albeit with a temporary change in structure and species composition and damage to some of the remaining trees. On the other hand, poor logging practices often cause much more damage to the forest than necessary, and deforestation may follow after the loggers have left, as landless people use logging roads to gain access to logged forest and clear it for agriculture – though this is not inevitable. Some environmentalists believe even selective logging has an unacceptable impact on the forests and many foresters also are concerned that logging often takes no heed of long-term sustainability and are calling for practices to be improved.

Deforestation and logging

So there are really two main human impacts on the tropical rain forests, deforestation and logging, although these are often confused in popular accounts. In this book we understand deforestation to refer to 'the temporary or permanent clearance of forest for agriculture or other purposes', which is in line with the definition used by the United Nations Food and Agriculture Organization (FAO) (Lanly, 1982). For deforestation to take place, according to this definition, forest has to be cleared so it can be replaced by another land use. This has the merit of requiring a clear physical change in vegetation cover that can be monitored by satellites and other remote sensing techniques.

Since most logging in tropical rain forest is selective, however, removing just a few trees per hectare rather than clearing the forest, it does not cause deforestation as defined here, though it does degrade the forest in varying degrees. Selective logging undoubtedly has a major impact on tropical rain forest but it is a very different one from deforestation and so deserves separate treatment. Focusing (sometimes exclusively) on trying to ban or control logging as if this would stop deforestation is therefore misguided. Indeed, as we shall see later in the book, it could have exactly the opposite effect. Different but complementary strategies are needed to control deforestation and improve forest management. Deforestation is more a land use problem than a forestry problem. An inability to recognize this led to the failure of previous attempts to control it. We will look at the definition of deforestation again in Chapter 1.

Sustainable land use

If deforestation is to be controlled farming must become more productive and sustainable overall. But this is easier said than done. Sustainable land uses are those that are capable of maintaining yields indefinitely without causing excessive environmental degradation. However, only limited areas in the humid tropics are suited to long-term intensive agriculture. The fairly even high temperatures and rainfall provide ideal growing conditions for trees and agricultural crops, but they have also resulted in mainly poor soils that have little inherent fertility and are vulnerable to erosion. So while clearance of temperate forests and continued cultivation of the soils beneath them actually improved the agricultural fertility of those soils this will rarely be the case in the humid tropics. Some areas are quite fertile and suited to sustained intensive farming, given good husbandry. But large areas are not suitable and there the only sustainable land uses are probably forest management, shifting cultivation, or other low intensity practices such as agroforestry. So those who criticize shifting cultivation or logging practices may be rather short sighted, for the only way to retain large areas of tropical rain forest could well be to encourage these land uses in less fertile areas.

Forest management must also become more sustainable in order to reduce the degradation caused by logging and give better protection from deforestation. But while the wealth of tree species in tropical rain forests, with thousands of species being found in individual countries, is something for ecologists to marvel at, it is a mixed blessing for foresters. Trying to harvest at predictable rates the timbers of the few marketable species, each of which usually occurs at a low density in the forest, presents a far greater challenge than that facing their counterparts in temperate countries. Loggers are expected to follow selective logging management systems specified by government foresters, but initially logging is more akin to 'mining' than managing the forests as if they were renewable resources, taking out valuable timber without giving much thought to future production. However good the management systems might be in principle, at the moment they are being poorly implemented, with widespread overlogging and excessive damage to remaining trees when commercial trees are felled and removed. In addition, because of the lack of previous scientific research into how tropical rain forest ecosystems function, both in their undisturbed state and after logging, today's selective logging systems are based on only a limited knowledge of how forests regenerate after disturbance, so their sustainability cannot be guaranteed even if they are implemented properly.

Development problems

On the basis of past experience it is unlikely that tropical agriculture and forestry can be made fully sustainable quickly. Farming and forestry, and land use patterns in general, all evolve as a society becomes more developed. 'Mining' timber without thought for tomorrow, for example, is not unique to the humid tropics; the same phenomenon was seen in temperate forests. In the USA it took three centuries before the need to manage them as renewable resources was appreciated. Are we right to expect tropical countries to pass through this transition within a mere decade or two? In less developed countries governments currently lack the organizational capacity and trained personnel necessary to control the management of large areas of lands and forests in remote areas, and they are often not subject to the kind of democratic controls that developed countries have found to be vital for proper environmental protection. Many tropical countries had their land use evolution biased by long periods of colonial rule in which cash crop plantations and resource exploitation became major economic activities. These newly-independent nations cannot simply abandon plantation production, logging and mining. They still need to earn foreign currency to fuel national economic development and it will take time for their economies to diversify.

Economic and social change lead to considerable pressures to exploit and clear forest. Agriculture expands in both planned and unplanned ways, relentlessly increasing its share of national land area at the expense of poorly protected forest as farm production rises more as a result of increased area than higher yields per hectare. The money needed to invest in more productive and sustainable practices is in desperately short supply, and the problem is exacerbated by widespread poverty, disparities in land ownership and other internal economic inequalities. Exploitation of timber and other natural resources is promoted in many countries more to benefit the short-term needs of the rich elite rather than long-term national interests. Eventually, as agricultural productivity increases and forest protection improves, we would expect a new balance to be struck between the proportions of national land area taken up by agriculture and forests, but it is likely that national forest cover will then be much lower than it was originally. The USA, for example, lost half its forests between the seventeenth and early twentieth centuries.

Global concerns versus local needs

So far we have taken a rather localized approach to what is after all a major global environmental problem. But the reason for this should be clear. At first sight we might assume that people all over the world would be eager to save the tropical rain forests for the benefit of humanity as a whole and that there is no reason in principle why this could not be done. Unfortunately, things are not so simple, because of two critical dilemmas. The first is to decide which deserves highest priority – people or resources? Should we conserve the tropical rain forests to protect for the whole world the vast natural riches they contain, or watch while they are cleared so that people in tropical countries can feed themselves and go through the same process of economic advancement that people in temperate countries now take for granted?

Secondly, the choice is not merely between tropical people and tropical resources, but between tropical people and temperate people. The latter have consumed many of their own resources and are now justifiably concerned to conserve those remaining on the planet as a whole, including the vast natural wealth found in the tropics. However, the vast majority of tropical rain forests are in developing countries and under their sovereign control. Environmental groups from the UK and USA caused much resentment in the 1970s when they appeared to be telling the governments of tropical countries such as Brazil what they should and should not do. This hectoring still continues today, though perhaps in a less strident fashion. Any country, industrialized or not, would get upset if lectured to by outsiders, but the resentment can get more intense in tropical countries that until relatively recently may have been ruled by one of the industrialized countries.

It is therefore important to find ways to conserve natural resources of global importance while at the same time recognizing the sovereignty of the countries in which the resources are found. The governments of those countries may quite reasonably object if asked to conserve large areas of tropical rain forests free from any exploitation without being given anything in return. Resource exploitation fuelled the development of the present industrialized countries so why should tropical countries be excluded from this, they may ask, especially as they are now so indebted to the developed world? Recognition of national sovereignty is one step towards solving this problem, another is finding how to value the benefits of the tropical rain forests fully so that the entire costs of foregoing exploitation, whether for global or national interests, can be included in national economic accounts and be fairly compensated by international financial transfers.

Monitoring and managing global environmental change

However much we might wish to understand the need for tropical deforestation, it is nevertheless occurring on a massive scale and is a major component of global environmental change. Humanity has now arrived at the chilling point in its evolution when it is able to modify not just local environments but the whole planet. In the developed countries carbon dioxide emissions from fossil fuel combustion are changing the composition of the atmosphere and threatening to change global climate through the greenhouse effect, while chlorofluorocarbons are depleting the planet's ozone layer. In the developing countries deforestation in the humid tropics in particular is also contributing to the greenhouse effect and reducing biodiversity, while poor land use in dry areas is causing desertification.

In these circumstances, now we have upset the natural balance of the planet in various ways, it is our duty both to monitor and manage our activities on a global scale. Yet here our conduct so far has been positively juvenile. Often the information available on global environmental problems such as tropical deforestation is sufficient to arouse concern but not reliable enough to support the development of policies that have a good chance of solving those problems. Estimates of remaining forest areas and deforestation rates in the humid tropics are very inaccurate, firstly because of the poor quality of national data, owing to the limited ability of tropical countries to monitor their own resources, and secondly because of the world community's lack of vision in devising ways to use satellite remote sensing technology, which has been at our disposal now for twenty years, to monitor the changes we are making to tropical rain forests and other resources. Many policy deliberations currently have to be conducted in a haze of ignorance for this reason. Effective global monitoring systems are needed, not just for tropical deforestation but for other global environmental problems too, such as desertification. It is also important to build up local monitoring capabilities, though initially these will probably provide data to serve mainly local needs such as better land use planning.

Managing the global environment, for that is literally what we must do if we are to bring our current depredations under control, needs both local and global solutions. The countries of the world must now move towards active cooperation on environmental protection. This does not just mean continuing the old system whereby governments planned national actions within a general international policy framework. The need now is for all countries to make binding commitments to achieve specific national targets so collectively they can fulfil global targets, e.g. reducing carbon dioxide emissions on an agreed timescale, and working out joint financing packages so that poorer countries can keep pace with richer ones in meeting their commitments. The world community's failure to agree at the UN Conference on Environment and Development (UNCED) at Rio de Janeiro in 1992 on binding targets to cut carbon dioxide emissions and on ways to fund this and other urgent programmes shows just how far we have to go if we are to become effective global environmental managers, instead of continuing to despoil the planet for short-term gain.

Global action is also constrained by the need for local solutions to complement internationally agreed policies. Tropical rain forest is not a homogeneous vegetation cover but consists of a variety of different types of ecosystems spread over three continents, varying in species composition from place to place. So we need to be selective in deciding which areas of forest to conserve if the fullest possible range of biodiversity is to be maintained. Furthermore, tropical deforestation cannot be cured by a 'quick fix'. Since it is caused by hundreds of millions of people acting largely independently, solving the problem requires working with them, even though a different approach may be needed in each area. Tropical rain forests cannot be safely protected in national parks, for example, without gaining the active cooperation of local people, and this takes time. The best way to change land use practices, of whatever kind, is usually to build on techniques that already work in each area, rather than trying to replace them. Traditional cultures are often far better at protecting forests than government organizations so it makes sense to try to ensure their survival.

Sustainable development – a shared experience

If any progress was made at UNCED perhaps it was the dawning of a realization by the governments represented there that all countries in the world are contributing to global environmental change, are likely to suffer from it in future, and are constrained in various ways in their ability to control it. Certainly, the discussion in these opening pages has suggested that people in temperate developed countries have very little right to criticize tropical countries about either deforestation or logging. Indeed, if they looked at their own history more closely they might understand current trends in the tropics better.

It is not just a question of morality, but of the changes that accompany social and economic development whenever and wherever it occurs. Whether or not development is 'good' or 'bad', we have to accept that it takes place. The questions that arise are how much deforestation is necessary for agriculture and how long it will take to move to more sustainable agriculture and forest management, and a new state of balance in national land use between agriculture and forests. By documenting the causes, extent and consequences of deforestation, describing current logging practices and the constraints on sustainability, and suggesting techniques and policies to facilitate the development of sustainable land management in the humid tropics, this book aims to provide a rational basis for answering these questions.

Left alone, countries in the humid tropics might well take at least another hundred years for this stage in their development, but that would not augur well for the tropical rain forests. Fortunately, the developed countries, while not renouncing their own development, are now beginning to view it in a different light, mainly because of the links between development, fossil fuel consumption and the greenhouse effect. There is increasing talk of the need for 'sustainable development', a pattern of development that broadly minimizes the depletion of natural resources and deterioration in environmental quality (World Commission on Environment and Development, 1987). As moves are made to define what this concept means and how it can be achieved, more aid should flow to tropical countries so they can make their development processes more sustainable. This should help to promote the positive changes in land and forest management needed to reduce the depredations of deforestation and logging. It will also demonstrate a recognition that *all* countries in the world now have a shared duty to move together as partners towards a more sustainable future. In this respect, aid from developed to developing countries should not be seen as flowing from teachers to students but just from richer partners to poorer partners, both of whom are still on a learning curve, albeit at different stages.

The plan of the book

This book looks at each of the major issues in turn. Chapter 1 provides some essential background material, looking at climate and soils in the humid tropics, the different types of tropical rain forests that occur there, tropical rain forest distribution, and the definition of deforestation. The next three chapters discuss how these forests are being exploited and why. Chapter 2 describes the direct causes of deforestation, including shifting agriculture, permanent agriculture and other land uses. Chapter 3 reviews selective logging practices, the issue of sustainability, and the historical development of the tropical hardwood trade and tropical rain forest logging. Chapter 4 looks at the underlying causes of deforestation, principally population growth and economic development and government policies, the reasons for the spread of logging and various aspects of tropical forest policy. Chapter 5 gives a critical review of estimates of deforestation rates. Chapter 6 assesses the possible impacts of deforestation on biological diversity, soil, water and climate. Chapter 7 outlines some agricultural, forestry, agroforestry and conservation techniques which could help to control deforestation. Chapter 8 suggests the changes to national and international policies needed to support the use of these techniques and tackle the underlying causes of the problem. Chapter 9 summarizes the main findings of the book, reviews progress made in controlling deforestation so far, and closes with some thoughts about future trends and prospects.

Controlling deforestation

There are no easy or quick solutions to tropical deforestation. To think otherwise would be counterproductive. This book argues that if we wish to bring the deforestation of tropical rain forests under control we must take an *integrated* approach to land use that encompasses agriculture, forestry, the management of other resources and conservation, rather than the more partial solutions proposed previously. Moreover, improved techniques by themselves are not sufficient to ensure a successful outcome. They must be backed by sound policies, and it is also important to address the underlying social, economic and political causes of the problem and the constraints that these impose on sustainable development in general.

This book presents *one* analysis of the problem as a contribution to the policy debate. There are other points of view, some of which are referred to at appropriate points in the text. Though comprehensive in scope the book is also intentionally brief, so it frequently has to generalize even though experiences differ from country to country, and often there is no space to discuss matters in the detail they deserve. The author hopes that the balance between brevity and detail is an acceptable one, and suggests that readers needing an expanded or more academic treatment consult another one of his books (Grainger, 1993).

Tropical rain forests and deforestation

Tropical rain forests thrive in the warm, wet environments of the humid tropics. The superb conditions for plant growth there allow a profusion of forests with a great diversity in species composition, but at the same time give rise to poor soils that make it difficult for farmers to capitalize on them. Farming can easily become unsustainable if practised too intensively on low quality deforested land. The great contribution that tropical rain forests make to the world's biological diversity is not just in their huge species content but in the variety of different types of tropical rain forest. The next section looks at each of the main types of rain forest in turn, and is followed by a review of the distribution of forests in the humid tropics. We lack estimates of the global and national areas of tropical rain forest and are forced instead to list areas of 'tropical moist forest' – the collective term for all closed forest in the humid tropics, including both tropical rain forest and the tropical moist deciduous forest found in seasonal areas. Crucial to the debate about tropical deforestation is the rate at which it occurs, but since estimates are influenced by the way in which deforestation is defined the final section of the chapter examines this matter and suggests the need to widen the usual definition, which requires forest clearance to encompass less severe human impacts.

Humid tropical environments

Climate and vegetation

The humid tropics forms a belt straddling the Equator that has high rainfall and moderately high temperatures throughout the year. The high rainfall, generally averaging 1,800 to 4,000 mm per annum and at least 1,200 mm, results from the ascent of warm moist air due to thermal convection and the meeting of the two sets of Trade Winds that flow towards the Equator from subtropical latitudes (30–40°N and S). The fairly even distribution of solar radiation during the year leads to constant high temperatures with little variation: mean monthly temperatures are generally 24–28°C and it never freezes.

The humid tropics extends over parts of the continents of Central and South America, Africa, Asia and Australia as well as islands in the South China Sea and Pacific Ocean. But it is only part of the wider tropical region located between the Tropics of Cancer and Capricorn (latitudes 23°N and 23°S respectively) and which includes dry as well as wet areas. Moving north and south from the Equator the climate becomes drier and more seasonal with increasing latitude, culminating in the two great desert belts that dominate subtropical latitudes where air flowing from the Equator eventually subsides, making rainfall formation difficult.

Nor is the humid tropics a homogeneous region. The simplest division is into permanently humid and seasonal zones. Tropical rain forest is found in the permanently humid tropics, the zone generally close to the Equator in which rainfall is distributed fairly uniformly throughout the year (Figure 1.1). The seasonal humid tropics may receive the same (or even more) annual rainfall but have a distinct dry season (60 mm of rainfall per month or less) that promotes the growth of tropical moist deciduous forest. This has a prominent component of leaf-shedding deciduous trees and is also known as 'monsoon forest', owing to the link in Asia in particular between rainfall seasonality and monsoon winds. Tropical rain forest and monsoon forest both owe their names to the German botanist Schimper (1898, 1903).

Soils

A long history of weathering and leaching has left soils in the humid tropics generally deep and devoid of nutrients. This may seem paradoxical given the profuse vegetation, but often a large proportion of nutrients in the ecosystem are contained in the vegetation rather than the soil[1]. Most tropical soils are deficient in phosphorus and susceptible in varying degrees to compaction, crusting and erosion (see Chapter 6). But not all soils are infertile, for there are small but significant areas of fertile soils, such as the alluvial and volcanic soils, that can support highly productive agriculture.

The world distribution of different types of soil is mapped using global classification systems such as the one devised by the US Department of Agriculture (USDA, 1975). This splits the world's

Tropical rain forests and deforestation

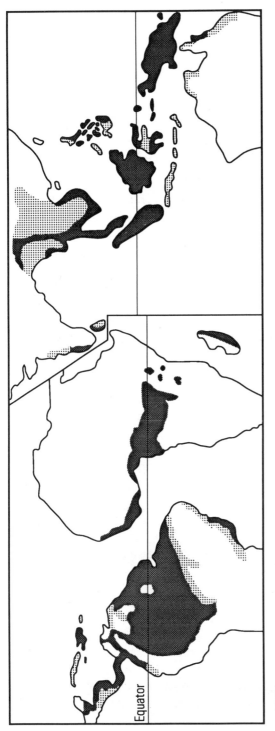

Equator

■ Tropical rain forest (includes tropical montane rain forest)

▦ Tropical moist deciduous forest (includes some subtropical rain forest)

NB. This map does not take account of deforestation
From various sources, including Eyre (1968)

Figure 1.1 *Distribution of tropical moist forest*

soils into ten major types, or orders, based on key soil characteristics (Table 1.1). Seven of the orders are found in the humid tropics but 63 per cent of the total land area is occupied by just two of them, the low fertility oxisols and ultisols (Figure 1.2) (Sanchez, 1976, 1981).

Table 1.1. *Tropical soils classified using the USDA soil taxonomy*

Order	Sub-orders	Characteristics
Orders present in the humid tropics		
Entisols	Aquent, fluvent, psamment	Immature, usually azonal
Inceptisols	Andept, aquept, tropept	Moderately developed
Alfisols	Aqualf, udalf	Moderate to high base content
Ultisols	Aqult, humult, udult	Low organic matter, low base supply
Oxisols	Aquox, humox, orthox	Nutrient poor, high in iron and aluminium, deeply weathered
Spodosols	Aquod, ferrod, humod, orthod	Accumulation of iron and aluminium oxides and/or organic matter under light sandy layer, podzols
Histosols	fibrist, folist, hemist, saprist	Peaty soils
Orders not present in the humid tropics		
Aridisols	–	Desert and semi-desert soils
Mollisols	–	Temperate grassland soils
Vertisols	–	Cracking clay soils with turbulence in profile

NB. Orders absent from humid tropics are not differentiated into sub-orders
Source: Goudie (1989), USDA (1975)

The oxisols and ultisols are red and red/yellow soils respectively, highly acidic, lacking in major plant nutrients, and with an excess of aluminium and manganese. Most soil nutrients are associated with soil organic matter, which is less abundant in ultisols than in oxisols. Ultisols are also more vulnerable to erosion when vegetation is removed. Oxisols are better drained but more susceptible to the formation of iron-rich subsoil concretions, called laterite, which restrict root growth. If these are exposed by soil erosion a hard brick-like

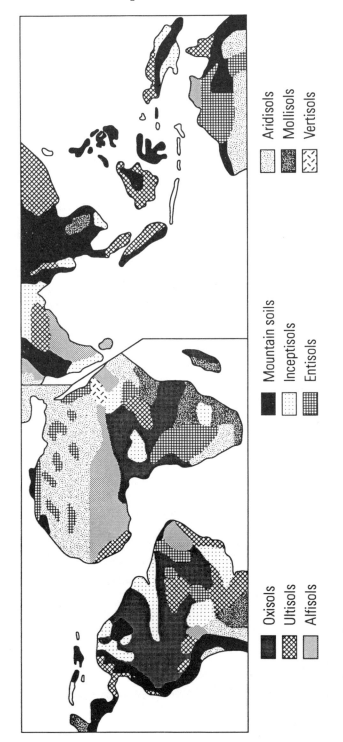

Oxisols
Ultisols
Alfisols

Mountain soils
Inceptisols
Entisols

Aridisols
Mollisols
Vertisols

Source: after Goudie (1989), USDA (1975)
NB. Mapping at this scale has major limitations and this map poorly represents the actual distribution of soil types in Southeast Asia, in particular.

Figure 1.2 *Distribution of soil orders in the tropics, according to the USDA Soil Taxonomy (Seventh Approximation)*

covering can form on the surface, but contrary to earlier expectations laterization threatens only a small proportion of all soils in the humid tropics, mostly in seasonal areas (Moorman and Van Wambeke, 1978). Oxisols and ultisols can be made more productive by adding lime (to reduce acidity) and nitrogen and phosphorus fertilizers. But these are expensive inputs for small farmers, need skilled application, and so are not usually a practical option at the moment.

The alfisols, richer in basic nutrients, are the next most common soil order but occur mainly in seasonal areas. Four other orders, the entisols, inceptisols, histosols (peaty soils) and spodosols (low fertility podzols) are found in more specialized environments, such as those close to rivers and on lands receiving regular inputs of volcanic ash. The andepts, a sub-order of the inceptisols formed from volcanic ash, are highly fertile and provide the basis for highly productive farming on the islands of Java, Bali and Sumatra (Indonesia) and Mindanao and Luzon (the Philippines). Older alluvial soils are also a type of inceptisol (aquepts) and distinguished from recently deposited alluvial soils classed as entisols (fluvents). Mountain soils are not easily classified by this scheme because soil type changes markedly with altitude.

Types of tropical moist forests

The tropics as a whole contains two main types of forests: closed forests, such as tropical rain forest and tropical moist deciduous forest, with a high density of trees and a closed canopy, and open forests, such as the savanna woodlands of the dry tropics in which trees are scattered at low density over grasslands. Closed forests do occur in the dry tropics but the majority are in the humid tropics. Tropical moist forests are mainly broad-leaved but small areas of coniferous (needle-leaved) and bamboo forests are found too. There are at least fourteen different types of tropical rain forest ecosystems (Whitmore, 1984, 1990), conveniently grouped here as dryland, wetland and montane rain forests (Table 1.2), as well as various types of tropical moist deciduous forest.

Forests are not the natural vegetation everywhere in the humid tropics, however. Savanna grasslands are found where soil conditions (poor drainage or nutrient content, a subsurface hardpan, etc.) are inhospitable to tree growth (Cole, 1986). They also occur in areas in Africa and elsewhere with a long history of clearance and burning for hunting and grazing. Asia has large areas of weedy grasslands (e.g. *Imperata* spp.) on sites where overcultivation and frequent burning by shifting cultivators has allowed them to become dominant.

Table 1.2. *Types of tropical moist forest ecosystems*

Site type	Forest type	Specific site/climate requirements
Dryland	Lowland evergreen rain forest	Everwet climate
	Lowland semi-evergreen rain forest	Short dry season
	Moist deciduous forests (various types)	Pronounced dry season
	Heath forest	Podzolized sandy soils
	Limestone forest	Over limestone
	Ultrabasic forest	Over ultrabasic rocks
Wetland	Freshwater swamp forest	
	Permanent swamp forest	Permanently wet
	Seasonal swamp forest	Periodically wet
	Peat swamp forest	Coastal or inland
	Saltwater swamp forest	
	Mangrove forest	Coastal
	Brackish water swamp forest	Coastal
	Beach forest	Coastal
Montane	Lower montane rain forest	From 700m^{WM}, 1500m^{NG}
	Upper montane rain forest	From 1500m^{WM}, 3000m^{NG}
	Subalpine forest	From 3000–3500m to treeline

NB. WM indicates approximate transition altitude in West Malesia, NG in New Guinea.
Source: adapted from Whitmore (1990)

Dryland rain forests

The group of dryland (as opposed to wetland or swamp) forests includes the richest of all tropical rain forest ecosystems: tropical lowland evergreen rain forest. This has the greatest height, biomass and species diversity and is fully evergreen. The closed canopy is 20–40 m high, and larger trees (called 'emergents') pierce the canopy to reach heights of 50 m.

The trunks of emergent trees often have huge buttresses (up to 7.5 m high), probably to provide additional mechanical support since the roots are quite shallow. Tree leaves are leathery to reduce moisture loss in the canopy and nutrient leaching by rainwater, and have pointed 'drip tips' to help water drain away quickly as a fine mist (thought to reduce splash erosion of soil when the droplets reach the ground). There are many herbaceous or woody climbing plants, rooting in the ground and growing as high as canopy trees but using trees for mechanical support. Orchids and other epiphytes also attach themselves to tree branches, either above or just below the rain forest canopy, and give the canopy a multi-coloured appearance. They do not root in the ground but collect nutrients from rainwater and dead organic matter.

Lowland evergreen rain forest requires a permanently humid climate. It is extensive in Southeast Asia, but only found in selected areas in South America, e.g. near the mouth of the River Amazon and the foothills of the Andes mountains and Guianas highlands, where rainfall is high and distributed uniformly through the year with minimum seasonality. It is probably absent from Africa. Evergreen rain forest in Southeast Asia is mainly dominated by trees of the Dipterocarp family, which have very valuable timbers.

In contrast, semi-evergreen rain forest is the dominant form of rain forest in Africa and Central Amazonia, and other areas in the humid tropics where rainfall is too seasonal to support evergreen rain forest. It contains fewer emergent trees and though up to a third of all trees are deciduous they do not all lose their leaves at the same time.

Of the other types of dryland rain forest, heath forest grows on poorly fertile acid podzolic soils. Its canopy is lower and more open than the two forest types above and notable for the many small saplings and pole-like trees. It occurs widely in Southeast Asia and also in Amazonia. Distinct forest types are also found on coastal beaches, and on basic soils, where limestone forest and ultrabasic forest are particularly notable (Whitmore, 1984).

Wetland rain forests

The profusion of rivers and wetlands in the humid tropics gives rise to a variety of swamp forests. Freshwater swamp forests grow at the interface between dryland forests and rivers. Those right next to rivers are permanently flooded, scrubby in form and the lower branches and roots of adjacent trees interweave. Further away from the river are the taller seasonal swamp forests that are only flooded for a few months each year. In Amazonia they contain valuable timber trees such as *Virola surinamensis* and *Cedrela odorata*, as well as the rubber tree *Hevea brasiliensis* and many palms.

Peat swamps are also covered by a distinctive forest type, similar in structure to heath forest but having species in common with both it and lowland evergreen rain forest. Peat swamp forest includes some very valuable industrial timber trees, most notably ramin (*Gonystylus bancanus*), found in Sarawak in western Borneo.

Other swamp forests grow in saltwater environments. Mangrove forests are still plentiful along coasts and estuaries in Southeast Asia but large areas have been cleared for agriculture in West Africa. Mangrove forest trees are prized for their charcoal and their characteristic features include stilt-like roots and many small tubes that rise above the water level to supply air to submerged roots. Another type of forest, in which *Nypa* palms are prominent in Asia, grows in seasonally flooded brackish water areas on the inland side of mangroves.

Montane rain forests

Mountains in the humid tropics show a distinct layering of different types of forest and vegetation types, from lowland evergreen rain forest at low altitude to alpine grasses, ferns and shrubs close to the summits. The reason for the layers is that as altitude increases the climate gets cooler and moister and the difference between daytime and evening temperatures widens, affecting the type of vegetation that can grow there. Between base and summit three main types of montane rain forests are found: lower montane forest, upper montane forest and subalpine forest, the first two being divided into sub-types according to local conditions. Montane rain forests are most abundant in Southeast Asia, where mountains occupy about 30 per cent of the total land area. Plateaux dominate the African and Latin American humid tropics so montane rain forests are more restricted there, e.g. to the slopes of the Andes and the eastern edge of the Congo Basin.

Lower montane rain forest takes over from lowland rain forest at about 1500 m above sea level on the island of New Guinea, which has the highest mountains in Southeast Asia and the western Pacific. The canopy is lower (30 m), there are fewer buttressed trees and climbing plants are absent. At about 3000 m upper montane forest begins, with a much lower canopy (6–21 m, and getting smaller with higher altitude), coniferous trees dominant, and tree trunks and branches covered by mosses and filmy ferns in the cloudy conditions. Higher up is the subalpine forest, with its small, gnarled trees only 4.0–7.5 m high and a profusion of rhododendrons. Alpine grasses, ferns and shrubs are the typical vegetation above the tree line at about 4000 m. The transition from one forest type to another varies with the overall height of the mountain: the smaller it is the more compressed the layers. So elsewhere in the region, where the mountains are not as high, lower montane forest is seen at lower altitudes.

Tropical moist deciduous forests

Tropical moist deciduous forest grows where there is a distinct dry season of up to four months, even though overall rainfall may exceed that in tropical rain forest areas. There are various types of tropical moist deciduous forest, just as with tropical rain forest, but generally the canopy is evergreen with deciduous emergent trees that lose their leaves in the dry season. Species diversity is lower than in tropical rain forest. Tropical moist deciduous forest covers large areas in continental Asia, including Thailand, Burma, Bangladesh, Nepal and India, and also spreads into insular Asia, including Java, Sulawesi, Luzon and Mindanao (Figure 1.1). The Asian monsoon forests contain teak (*Tectona grandis*), a valuable timber tree, and pines as well. A band of tropical moist deciduous forest is found between tropical rain forest and savanna in parts of South America, but usually not in Africa where the transition is more abrupt. It grows in the west of Central America while tropical rain forest covers the eastern side. Its contribution to current deforestation rates is probably low because large areas were cleared long ago: the seasonal climate and superior soils made the lands very attractive for settlement and agriculture and still today they are regarded as better locations for timber plantations than permanently humid areas.

Tropical rain forests – flourishing yet vulnerable

To those privileged to walk inside them tropical rain forests are likely to instil a sense of awe and wonder that such a marvellous assemblage of plants and animals can exist. But eventually curiosity, another great human virtue, rears its head. Given the limitations of humid tropical environments how on earth do tropical rain forests manage to flourish there?

Over the last century tropical ecologists have been grappling with this intriguing question. The answer has much to do with the structure and functioning of the ecosystems, though our knowledge is still quite limited, and there is only space here to highlight some key findings that are explored in more detail in other books (Jordan, 1985; Longman and Jeník, 1987; Mabberley, 1992, Whitmore, 1984, 1990). Despite the excellent growing conditions, for example, most trees in tropical rain forest do not grow continuously (Longman and Jeník, 1987). Nor, for that matter, do they include the world's tallest or oldest trees. On the other hand, the tightly packed collection of vegetation results in a dense canopy that is very efficient at collecting solar radiation and converting it into biomass.

Another characteristic is the great diversity of plant and animal species. Contributing to this are the complex vertical and horizontal structures of the plant community, the way that forest microclimate changes from the canopy to the forest floor, and the many interdependent relationships between plant and animal species, all of which create numerous niches for plants and animals. Interdependence between plants and animals is particularly evident in the reproductive mechanisms of some tree species, which depend on a small number of animals for flower pollination and seed dispersal. When added to the widespread irregularity in fruiting and flowering, a general lack of seed dormancy and other factors, such mechanisms constrain the ability of the forest to regenerate after logging or clearance and are a major challenge to foresters keen to improve the sustainability of tropical rain forest management for timber production (Longman and Jeník, 1987; Whitmore, 1990).

Tropical rain forest ecosystems are able to flourish on generally poor soils because of their nutrient storage and cycling mechanisms. It was previously assumed that most nutrients were stored in the vegetation rather than the soil, in order to protect the nutrients from being leached by the heavy rain, and that nutrient cycling was rapid and extremely efficient. But recent research has shown this to be an over-generalization and just one extreme of a range of characteristics seen in different ecosystems. Soil type has a key influence: forests on poor soils do tend to store most nutrients in the vegetation and recycle nutrients with minimum leakage to or input from the wider environment, but the situation is different in forests on better soils. The rate of cycling is also generally slower in forests on less fertile soils. The old model was therefore over-simplistic, but it is still useful, for by emphasizing the need to maintain a continuous vegetation cover to sustain the fertility of the land as a whole it provides a good basis for designing sustainable farming systems (Jordan, 1985; Proctor, 1989).

The distribution of tropical rain forest

Tropical rain forests occur throughout the humid tropics but despite overall similarities the forests in Southeast Asia and the Pacific, Latin America and Africa are not one vast homogeneous carpet of greenery. They differ so much in their dominant families, genera and species of trees and the relative proportions of the different forest sub-types that the three regional blocks, or 'biomes', that make up the tropical rain forest 'biome type' are quite distinct (Whittaker, 1975). Almost three fifths of all tropical moist forest is in Latin America, a quarter in Asia and the Pacific, and under a fifth in Africa.

The three biomes

The Asian (Indo-Malayan) biome contains a large area of tropical lowland evergreen rain forest. It is centred on the 'Malesian' archipelago – a floristic region that includes Malaysia, Indonesia, the Philippines and Papua New Guinea – and extends northwards into continental Asia (Thailand, Burma, Vietnam and India) and eastwards to Australia and the Pacific islands (hence the regional term 'Asia-Pacific') (Figure 1.1). North of Malaysia's border with Thailand the forests become increasingly deciduous, grading into semi-evergreen rain forest before tropical moist deciduous forest becomes dominant. The Indonesian islands of Java, Bali, Sumbawa, Flores, Timor, Sulawesi, etc. have a seasonal climate and here tropical moist deciduous forest is the major climax vegetation. This splits the biome into two parts, the west being more species-rich than the east.

The American biome is centred on the basins of the rivers Amazon and Orinoco. Much of central Amazonia is covered by semi-evergreen rain forest. Evergreen rain forest is essentially restricted to areas near the mouth of the River Amazon and the Andes and Guianas foothills (Whitmore, 1984). Brazil has the largest area of tropical rain forest in the region (and the world) but there are substantial tracts in Colombia, Venezuela, Bolivia, Peru, Guyana, and Ecuador too. Colombia and Ecuador have rain forest on their narrow Pacific coastal plains as well but much of this has been cleared; the same is true for the strip of tropical and subtropical rain forest on Brazil's Atlantic coast (Mori, 1989). Tropical rain forest is still found on some Caribbean islands and in eastern Central America, particularly Costa Rica, Nicaragua, Honduras and Mexico. Western Central America is drier and covered with tropical moist deciduous forest and savanna. The Latin American region generally refers in this book to South America, Central America and the Caribbean islands.

Most of Africa's remaining tropical rain forest is in Central Africa, centred on the Congo Basin. West Africa once had a substantial area, stretching from Nigeria to Sierra Leone, split by a strip of dry forest in Benin, Togo and much of Ghana, with gallery, mangrove and other moist forest fragments as far north as Senegal, but much of this has been cleared or degraded. Lowland rain forest extends eastwards into Uganda but elsewhere in East Africa, apart from Madagascar, distribution is patchy and restricted to montane areas in Kenya, Tanzania, Rwanda and Burundi and limited areas on the islands of Mauritius, Reunion and the Seychelles (Polhill, 1989; Lanly, 1981). The African biome is the poorest of the three in composition and structure. Evergreen rain forest is thought to be absent (Whitmore, 1984).

Total forest area

Before human beings arrived on the scene, tropical moist forest is thought to have covered about 1603 million ha (Sommer, 1976). Probably less than 1000 million ha of this remains today, representing almost two fifths of all the world's closed forests. If collected together it would cover an area greater than that of the USA. Tropical rain forest could account for about two thirds of all tropical moist forest (Persson, 1974).

Adrian Sommer, in a report for FAO, estimated the area of tropical moist forest specifically for the first time in the 1970s as 935 million ha (Table 1.3) in the 65 countries that he listed as comprising the humid tropics. Comparing this with later estimates is difficult as he only gave regional, not national forest areas. The present author used data from Reidar Persson (1974) and Norman Myers (1980) to arrive at similar estimates of 979 million ha[2] and 972 million ha[3] respectively (Grainger, 1983, 1984). FAO and Norman Myers have made most of the tropical forest area estimates since 1970. Other estimates are mainly secondary ones, derived from FAO or Myers rather than new data, and are usually of limited value.

Table 1.3. *Estimates of tropical moist forest area (million ha)*

Source	Date	Area
Persson (1974)[a]	1970s	979
Sommer (1976)	1970s	935
Myers (1980)[a]	1970s	972
Grainger (1983)	1980	1,081
Myers (1989)	1989	800
FAO (1990)	1990	1,282

[a]Derived from source data by Grainger (1984)

The first really thorough survey of tropical forests was the joint 1980 FAO/UNEP Tropical Forest Resources Assessment Project (Lanly, 1981), prepared by FAO under the auspices of the Global Environmental Monitoring System of the UN Environment Programme (UNEP). This gave the total area of all tropical closed forest (i.e. both moist *and* dry) as 1,201 million ha (Table 1.4). Data in that report were used by this author to estimate that *tropical moist forest* alone covered 1,081 million ha in 1980 in 61 of the 65 countries listed by Sommer

(1976) as well as Puerto Rico and Vanuatu (Tables 1.5 and 1.6) (Grainger, 1983, 1984)[4].

Table 1.4. *Areas of tropical closed forest, tropical moist closed forest, tropical dry open woodland and all tropical forest in 1980 (million ha)*

	Closed Forest		Open Woodland[1]	All Tropical Forest
	All[1]	Moist[2]		
Africa	217	205	486	703
Asia-Pacific	306	264	31	337
Latin America	679	613	217	896
Total	*1,201	*1,081	734	1,935

NB. *Totals are not the sum of regional figures owing to rounding.
Sources: 1: Lanly (1981); 2: Grainger (1983), based on Lanly (1981);

Table 1.5. *Estimates of regional tropical moist forest areas (million ha)*

	Persson 1970s[2]	Sommer 1970s[2]	Myers 1970s[1]	Grainger 1980	Myers 1989	FAO 1990[4]
Africa	183	175	156	205	152	405
Asia-Pacific	249	254	290	264	211	190
Latin America	547	506	527	613	416	687
Total[5]	979	935	973	1,081	800[3]	1,282

1: Derived from source data (Persson, 1974; Myers, 1980) by Grainger (1984)
2: Sommer (1976)
3: Regional figures are totals of individual countries listed by Myers (1989). The total was adjusted by him for the whole biome-type
4: FAO (1990), revised for UN Conference on Environment and Development (1992)
5: Totals may not always equal sum of regional figures due to rounding

Table 1.6. *Estimates of national tropical moist forest areas in 1980 and 1989 (million ha)*

	Grainger 1980	Myers 1989
Africa		
Cameroon	17.9	16.4
Central African Republic	3.6	na
Congo	21.3	9.0
Equatorial Guinea	1.3	na

Tropical rain forests and deforestation

Gabon	20.5	20.0
Ghana	1.7	na
Guinea	2.1	na
Guinea-Bissau	0.7	na
Ivory Coast	4.5	1.6
Liberia	2.0	na
Madagascar	10.3	2.4
Nigeria	6.0	2.8
Sierra Leone	0.7	na
Uganda	0.8	na
Zaire	105.7	100.0
Asia-Pacific		
Brunei	0.3	na
Burma	31.2	24.5
India	16.7	16.5
Indonesia	113.6	86.0
Kampuchea	7.2	6.7
Laos	7.6	6.8
Malaysia	21.0	15.7
Papua New Guinea	33.7	36.0
Philippines	9.3	5.0
Thailand	8.1	7.4
Vietnam	7.4	6.0
Latin America		
Belize	1.3	b
Bolivia	44.0	7.0
Brazil	331.8	220.0
Colombia	46.4	27.8
Costa Rica	1.6	b
Cuba	1.3	na
Dominican Republic	0.4	na
Ecuador	14.2	7.6
El Salvador	0.1	b
French Guiana	8.9	a
Guatemala	3.8	b
Guyana	18.5	a
Honduras	1.9	b
Mexico	14.8	16.6
Nicaragua	4.2	b
Panama	4.2	b
Peru	68.8	51.5
Surinam	14.8	a
Venezuela	31.9	35.0

a:Total forest area for French Guiana, Guyana and Surinam given as 41 million ha (cf 42.2 million ha in 1980). b:Total forest area for Belize, El Salvador, Honduras, Nicaragua, Costa Rica and Panama given as 9 million ha (cf 13.3 million ha in 1980). na indicates data not available.

Sources: Grainger (1983), Myers (1989).

It was surprising to find that the forest area in 1980 was higher than stated in earlier estimates, as considerable deforestation had undoubtedly taken place in the 1970s, but understandable given the general inaccuracy of global estimates. It is best to put the difference to one side and regard the FAO/UNEP report as the first of a new generation of more accurate forest assessments. It listed national forest areas for 76 countries, divided by biological type (broad- leaved, coniferous, closed, open, etc.) and timber production potential (logged, unlogged, managed, unproductive, etc.); used data from remote sensing surveys of various countries in the 1970s; cited in most cases how and when areas were surveyed; corrected all estimates to the same year (1980), no matter when the latest survey had been made; and ensured homogeneity between countries by categorizing raw data with a uniform system of forest classification – previous assessments had listed areas for different types of forests.

Despite these advances the accuracy of the FAO/UNEP data was still not very high. Less than three fifths of all tropical moist forest had been surveyed since 1970 by any form of remote sensing. This included Landsat surveys of 19 countries (12 undertaken specially for the project), side-looking airborne radar (SLAR) surveys of four countries, and aerial photographic surveys of five countries together with parts of Malaysia. Brazil's national SLAR survey in the 1970s accounted for half of the surveyed area (Grainger, 1984). Any gaps were filled by FAO staff, using data from government forest departments or other sources.

There have been four more recent estimates. In 1989 a second report by Norman Myers claimed that tropical moist forest area was now down to 800 million ha. This was based on a total of 778 million ha for 27 individual countries and a group of 7 countries in Central America, adjusted upwards by Myers to take account of the 3 per cent of total forest area outside these countries (Tables 1.5 and 1.6). While this might seem the most 'up-to-date' estimate Myers was less rigorous than FAO/UNEP, making many assumptions which, while mentioned in the text, are difficult to substantiate. He listed the data sources on which his estimates were based but gave little information to show whether they came from subjective expert judgements or actual measurements, and if so which ones and when they were undertaken. Just four countries were stated to have declined in forest area by a total of 173 million ha compared with the 1980 figures in Grainger (1983): Brazil, by 110 million ha; Indonesia, by 28 million ha; Colombia, by 18 million ha; and Peru, by 17 million ha. If the actual reductions were lower then Myers' estimate would be more comparable with the latest FAO figures.

FAO was preparing its next tropical forest assessment as this book was being completed but has released three interim estimates. The first put the area of tropical moist forest in 1990 at 1282 million ha – even larger than the area of *all* tropical closed forest in 1980 (FAO, 1990). The second interim estimate in 1991 gave 1714 million ha as the area of all tropical forest in 87 countries in 1990 (FAO, 1991). This was comparable with FAO/UNEP's estimate of 1935 million ha for all open and closed forest in the dry and humid tropics in 1980[5]. The third estimate, presented to the UN Conference on Environment and Development in 1992, was more disaggregated than the second but retained 1282 million ha as the area of tropical moist forest (Table 1.5) and included an estimate of 655 million ha for the area of tropical rain forest. The 1282 million ha total seems very high – an area between 900 and 1000 million ha would be more likely – and the figure for Africa is significantly greater than that found in earlier estimates. However, it is probably best to reserve judgement until the final report of FAO's 1990 Tropical Forest Resource Assessment Project is published

The accuracy of estimates of tropical moist forest areas clearly leaves much to be desired. This stems mainly from inadequate national surveys and the lack of a proper global monitoring programme capable of using available satellite data. Global estimates still depend heavily on government statistics, but national forest surveys on which they are based have rarely been carried out more frequently than once every ten years, with longer gaps in many instances.

The leading forest countries

Disregarding the inaccuracy of the figures at our disposal, we see that over a half of all tropical moist forest is in Latin America, a quarter in Asia and the Pacific, and less than a fifth in Africa. Three countries account for 55 per cent of all tropical moist forest: Zaire, Indonesia, and Brazil – which alone has 30 per cent (Table 1.6). Peru, Colombia and Bolivia contain a further 15 per cent. South America has 95 per cent of all tropical moist forest in Latin America, with only 5 per cent in Central America and the Caribbean. In Asia-Pacific, Indonesia contains more than two fifths of all forest in the region, Papua New Guinea and Burma together a further quarter, and Malaysia less than a tenth. Four fifths of all tropical moist forest in Africa is in just four countries in Central Africa: Cameroon, Congo, Gabon, and Zaire – which alone accounts for a half. West Africa has less than a tenth of the African total.

Controlling tropical deforestation

What is deforestation?

Lack of monitoring also leads to poor estimates of deforestation rates, as will be seen in Chapter 5. Deforestation is defined in this book as the temporary or permanent clearance of forest for agriculture or other purposes. This is compatible with FAO's definition (Lanly, 1982). The key word here is clearance: if forest is not cleared then deforestation does not take place. Often, but not always, forest is cleared so it can be replaced by another land use, such as cattle ranching.

But there are other human impacts on tropical rain forest besides clearance for agriculture and mining, etc. As explained in the Introduction, although clear-felling, the clearance and removal of all trees in a stand, would constitute deforestation, it is rare in the humid tropics where the common forestry practice is selective logging. This does not cause deforestation as defined here since it removes only a few trees per hectare and leaves the rest behind. However, some trees are damaged, forest structure is temporarily changed when the canopy is disturbed and there are also changes in biomass and species composition. Many of these changes will be reversed over time as forest regenerates but they are changes none the less.

Just because we adopt a definition of deforestation that is clear in a physical sense and convenient for measurement it does not mean we should ignore selective logging and other lesser impacts. To the conservation purist who sees any human incursion into tropical rain forest as detrimental, whether it causes clearance or not, the differences between logging and deforestation are not important, but to those concerned with the specific effects of human impacts, e.g. on biodiversity and global climate, they are absolutely crucial. That is why we need to widen the scope of our discussion to define these impacts as well as we can.

A good way to do this is to speak of *degradation*, which we can define as a 'temporary or permanent deterioration in the density or structure of vegetation cover or its species composition'. This encompasses the effects of selective logging, which causes a temporary change in canopy cover, forest structure, biomass and species composition. Further illegal logging by poachers after concessionaires have left will cause more degradation. But after any deforestation event, some form of vegetation should return to the land as crops, pasture or regrowth, although in its density (which includes both biomass and canopy cover), structure and species composition, it will usually be inferior to the forest it replaces and so degradation has occurred. Deforestation, in which the density of vegetation cover is temporarily reduced to zero, is obviously the most extreme form of degradation.

The term degradation is useful in accounting for the effects of short-rotation shifting cultivation (see Chapter 2), in which secondary forest (forest fallow) grows on abandoned plots but never reaches the same biomass or overall quality as mature forest since it is cleared again for cropping after a few years. Forest fallow is clearly degraded relative to the primary forest that covered the land before. The clearance of forest fallow before the next cropping period has been called deforestation by some commentators, but it is not the same as the clearance of forest which has never been cleared or has reached a high degree of maturity since the last clearance. Regular regrowth and clearance are a normal part of shifting cultivation so it is misleading to refer to subsequent clearances as deforestation.

Degradation can also be used to refer to the effects of human impacts over time, in contrast to the event-based terms of 'deforestation' and 'logging'. Selective logging probably has both short- and long-term effects on biodiversity, though they are poorly understood. Degradation is also caused by over-hunting and pollution of air and water, and as short-rotation shifting cultivation spreads into an area and forest fallow becomes the typical woody vegetation in place of primary forest, a long-term process of degradation occurs. Scientists studying the role of tropical rain forests in global climate change need to monitor degradation because it plays a crucial role in the transfer of carbon between biomass and atmosphere (see Chapter 6).

It should be noted, in connection with long-term degradation, that ecologists are now very cautious about labelling forest as either 'virgin' – that which has never been disturbed by human beings – or 'primary' – which may have been disturbed long ago but has since undergone a process of succession through different types of 'secondary' forest before regaining the structure and species composition characteristic of the climax ecosystem, a process that can take a thousand years or more. This caution derives from the finding of archaeological remains in various areas of what were previously thought to be primary forest, and the realization that human settlement in the forests of West Africa and Central America in particular has a long history. The mahogany tree, *Swietenia macrophylla*, from Latin America, for example, is actually characteristic of secondary, not primary forest. Some secondary forest is very old indeed, but while it is 'mature' it is not 'primary'. David Mabberley, of the Department of Plant Science, University of Oxford, has summed up the matter thus: 'most forests are probably "disturbed" but some are more disturbed than others' (Mabberley, 1992).

Although 'deforestation' will probably continue to be used to refer to forest clearance it could be used in parallel with 'degradation', or a similar term. But it is much more difficult to measure degradation by remote sensing techniques so some time could elapse before reliable numerical estimates of degradation rates are available, especially bearing in mind that the full potential of these techniques for monitoring deforestation has still not been exploited.

Forests, tropical environments and deforestation

Tropical rain forests present us with many paradoxes. They demand a uniformity of climate yet are the most diverse ecosystems in the world. They have the highest biomass per unit area but grow on some of the poorest soils, which are for the most part unsuited to intensive agriculture. The great green blanket of forest, speckled with the rich colours of orchids, may appear endless and homogeneous from the air, but actually comprises a variety of different types of tropical rain forest. There is no space here to talk of the structure and functioning of these majestic ecosystems in anything but a cursory fashion but when they are cleared most of the mechanisms that enable them to flourish disappear.

It is, if you think about it, tremendously impressive that human beings in the temperate world can show concern for exotic tropical forests outside their own regions that cover a significant portion of the globe. The next three chapters challenge us to show concern also for our fellow human beings who live in the tropics, and to understand the problems they face in trying to manage their environments and improve their lives. If these two concerns can be combined we shall have a good basis for controlling deforestation.

2

The causes of deforestation

Tropical deforestation is more a land use problem than merely a forestry problem and it usually takes place so that forest can be replaced by another land use. A larger number of land uses are involved than is generally supposed, though they vary in dominance from one country to another and also differ in their impacts on the forest. This chapter looks in turn at each of the main land uses, dividing them for convenience into three groups: shifting agriculture; permanent agriculture, including such practices as wet rice cultivation and tree crop plantations; and other land uses such as mining and hydroelectric power schemes (Table 2.1). Since their introduction provides the motive for forest clearance land uses may be considered to be the direct, or immediate causes of deforestation. But to understand *why* it occurs we must look to some deeper underlying causes, including population growth, economic development and government policies. They are mentioned here but analysed in greater detail in Chapter 4.

Shifting agriculture

Shifting, or slash-and-burn, cultivation has historically been the first stage in agricultural development worldwide when people make the transition from hunter-gathering, which entails the hunting of animals and the gathering of fruits and other plant parts. Shifting cultivators now comprise the majority of forest dwellers, an estimated 200-300 million people. Most hunt and gather in the forest as well, but there are only 1 million pure hunter-gatherers, restricted to a few isolated tribal peoples in the Congo Basin, Amazon Basin, Borneo and elsewhere (World Bank, 1978; Caufield, 1984). Traditional shifting cultivation is one of the few proven sustainable land uses in the humid tropics, but the two other main types described here, short-rotation shifting cultivation and encroaching cultivation, are less sustainable[1].

Table 2.1. Land uses which replace forest after deforestation

Agricultural land uses

1. Shifting agriculture
    a. Traditional long-rotation shifting cultivation
    b. Short-rotation shifting cultivation
    c. Encroaching cultivation
2. Permanent agriculture
    a. Permanent staple crop cultivation
    b. Fish farming
    c. Government sponsored resettlement schemes
    d. Cattle ranching
    e. Tree crop and other cash crop plantations

Other land uses

3. Mining
4. Hydro-electric schemes
5. Cultivation of illegal narcotics

Traditional shifting cultivation

Shifting cultivation is by definition an itinerant practice that involves clearing a small patch of forest (typically 1–2 hectares per family), burning the vegetation so nutrients are freed (as ash) and washed by rain into the soil, and then growing crops on the land for one to two years. The site is then abandoned and cultivation shifts to a different site, where another patch of forest will be cleared. The first patch will not be used again until it has been under fallow for some years. Although yields of successive crops decline the main reason why farmers move to a new site is often the growth of weeds.

In the traditional practice the fallow period between successive croppings of the same patch of land is at least 15 to 20 years and often longer. This allows forest to regenerate and the fertility of the land to be restored as the nutrient contents of both vegetation and soils are replenished. Regeneration also protects the soil from erosion and controls the spread of weeds, pests and diseases. Traditional shifting cultivation is an extensive practice but poses little threat to the integrity of forested areas. The emphasis is on the subsistence cultivation of staple crops such as rice, maize and cassava, the choice of crops varying from country to country. Quite complex systems have been devised in which a number of crops are grown together or sequentially.

One drawback of traditional shifting cultivation is that it can only support low population densities, typically less than 20 people per square kilometre, and consequently it is now found only in fairly remote areas where enough land is still available for such an extensive practice. Although its future is in doubt, it could survive well into the next century if the spread of permanent agriculture onto poor quality land is constrained by better land use planning or access to forested areas is restricted by government .

Short-rotation shifting cultivation

Most shifting cultivators now employ shorter rotations than the traditional system, with fallow periods of typically 6 to 15 years and even less in some places. With the rise of populations and encroachment of commercial land uses into shifting cultivation areas there is simply not enough land to support long rotations any more. The nomadic lifestyle has also become unattractive, as people aspire to a more settled village-based existence that allows them increasing access to the advantages of 'modern society'.

But shorter fallow periods give forest less time to regenerate, and this results in large areas of low bushy regrowth, called 'forest fallow', that is quite distinct from primary or old secondary forest. There is less time also for fertility to be replenished, and this carries with it a risk of declining yields, soil erosion, and general unsustainability. The minimum fallow needed for long-term sustainability differs from place to place and with the cropping system employed (Young and Wright, 1980). As the frequency of clearing and burning increases so too does the risk of invasion by weeds, such as alang-alang (*Imperata cylindrica*), which spread by underground stems, are fire resistant and inhibit forest regrowth when they become dominant. Alang-alang covers an estimated 14 million ha in Kalimantan (Indonesian Borneo) and up to 20 per cent of formerly forested land in the Philippines (Suryatna and McIntosh, 1980; Kartawinata, 1979).

Encroaching cultivation

Whatever its faults, short-rotation shifting cultivation is usually practised with the aim of sustainability. This is not true of encroaching cultivation (Eckholm, 1976, 1979), also called 'shifting cultivation by necessity' (Watters, 1960). It is typically carried out by migrant landless people who use newly-built roads to gain access to forested

areas, spreading out from roads in waves, clearing forest, and cultivating the land for three to four years or more until fertility has been
depleted, weed infestation is severe and crop yields are low. Only
then do they move on to clear another patch of forest, leaving behind
a devastated wasteland covered by bush, weeds and scrub on which
forest does not easily regenerate. This does not concern them very
much as they usually have no intention of returning. Encroaching
cultivation results in forest clearance on a much larger scale than the
other forms of shifting cultivation, so its overall impact is similar to
that of permanent agriculture. The area of degraded, deforested land
abandoned by shifting cultivators every year in the Philippines
alone in the early 1980s was officially estimated at 200,000 ha but
could have been far higher (Myers, 1985a).

With an estimated 800 million landless people in the tropics
(Myers, 1985a) encroaching cultivation is a serious problem, but
remarkably little is known about it. Many encroaching cultivators
are not skilled farmers and simply clear land to grow food to survive.
Others may have previously been permanent cultivators, forced
from their lands by poverty or other reasons. Another group will
have been paid to clear forest by outside agents (e.g. speculators or
ranchers) to establish rights over that land so they can use it later.
Encroaching cultivation is not necessarily a pure subsistence practice, and may involve the cultivation of cash crops, such as pepper
in Kalimantan and opium poppies in Thailand (UNESCO, 1983a,
1983b).

Encroaching cultivation is often a response to poverty and
government inaction over land reform, and indirectly facilitated by
government-backed schemes that build highways or logging roads
into previously inaccessible forested areas. Inequitable systems of
land tenure, in which a large proportion of the land in a country is
owned by a small number of families, are a particular problem in
Latin America and the Philippines, for reasons that date back to the
time of Spanish and Portuguese colonial rule hundreds of years ago.
Thus in Brazil 42 per cent of all cultivated land is owned by a mere
1 per cent of all farmers (Myers, 1985a).

The grand promises of politicians to redress this imbalance usually come to nought. President Marcos launched a land redistribution policy in the Philippines in 1972 but it was not very successful.
Fifteen years later President Aquino's plan to split up large estates
also ran into problems (Gourlay, 1987a, 1987b, 1987c, 1987d, 1987e).
In Brazil, President Sarney finally promised in 1985 to implement a
1976 land reform law but little was done. Although land reform is
vital to curb deforestation caused by encroaching cultivation it could

lead to more deforestation. Many large landowners in Brazilian Amazonia burned part of the forest remaining on their properties in the 1980s to try to show they were making 'productive use' of the land so it would not be taken away from them by an intended land reform programme (Charters, 1985).

Landlessness is becoming more serious every year owing to growing populations, changes in agricultural practices reducing the need for agricultural labourers on large estates, declining food yields in overcrowded rural areas (such as Brazil's degraded and drought-prone north-east region), and the spread of poverty. Landless people may move to cities in search of work but if they are unsuccessful they may migrate to rural areas where land is thought to be available and become encroaching cultivators.

When the start of the TransAmazonian Highway programme was announced by the Brazilian government in 1970, linking Recife on the north-east coast with Cruzeiro do Sul in the state of Acre, near the border with Peru, it was claimed that it would provide a channel by which poverty-stricken people from north-east Brazil could colonize Amazonia through government resettlement schemes. A bad drought in that year had brought the problems of the north-east to public attention, but this initiative was actually a continuation of a scheme begun in the late 1950s to promote settlement and economic development in Amazonia by building new roads (Figure 2.1). The first highway (BR 010), between Belém at the mouth of the River Amazon and the new federal capital, Brasilia, was completed in 1960 and paved in 1964. It was followed by BR 364 between Cuiába in Mato Grosso state and Porto Velho, the capital of Rondônia, which was finished in 1964. The first stage of the TransAmazonian Highway, linking Estreito (on BR 010) with Itaituba was completed in 1972 and the second, between Itaituba and Humaitá, in 1974. The final stage, between Humaitá and Cruzeiro do Sul, was never built.

These and other highways have since been used by millions of official and spontaneous migrants, leading to an annual population growth rate of 6 per cent in Amazonia as a whole during the 1970s. The TransAmazonian Highway scheme never realized its objectives: most settlers came from the south-east of the country rather than the north-east, and many of them eventually returned home when they discovered how poor the land was (see below) (Mahar, 1989). By far the greatest migration into Amazonia has been from the south-east of Brazil along BR 364, quadrupling the population of Rondônia between 1970 and 1980 (Fearnside, 1985a).

Permanent agriculture

Permanent, as opposed to shifting, agriculture is practised in the humid tropics as in temperate countries, but with some differences. Grain crops are grown in permanent fields but in the case of rice these are usually flooded. Cattle are grazed on permanent pastures, but this is rare in Africa due to the sleeping sickness disease spread by the tsetse fly, and more highly developed in Latin America than Asia. A distinctive form of tropical agriculture is the plantation of perennial tree crops like rubber. Permanent agriculture is now expanding to keep up with growing demand for food and foreign currency. But land suitability is always a constraint for, as Chapter 1 showed, only a small proportion of land, mainly that covered by alluvial and volcanic soils, is fertile enough to support continuous intensive farming.

Staple food crops

Permanent cultivation of staple food crops by both small and large farmers takes place throughout the humid tropics, with rice, cassava and maize cultivated extensively (but not exclusively) in Asia, Africa and Latin America respectively. Rice is the world's second most important cereal crop after wheat. Its cultivation in flooded fields (or paddies) is common in Asia on alluvial soils in river valleys and coastal plains and on other fertile flat lowland sites that are easily flooded and well drained (Grigg, 1977). It is usually sustainable in these conditions and supports higher population densities than cassava and maize (World Bank, 1982). The flooded conditions reduce soil erosion and soil acidity, and extra nutrients are imported in the water through nitrogen fixation by algae and wastes such as water buffalo manure (Murton, 1980). Farmers practise rotational fallowing of paddy plots.

Rice yields under irrigated conditions are higher than under shifting cultivation and even higher yields were made possible by the development of high-yielding varieties during the 'Green Revolution' in the Philippines and other countries in the 1960s and 1970s. But because seeds, fertilizers and pesticides are expensive for small farmers the Green Revolution tended to benefit only the larger farmers (Gollin, 1987; Pierce, 1990). Daniel K. Ludwig introduced highly mechanized wet rice cultivation in his massive Jari project in Brazilian Amazonia but it was not economically viable, owing to poor soil fertility and insect infestation (Fearnside, 1985b; Anon, 1987).

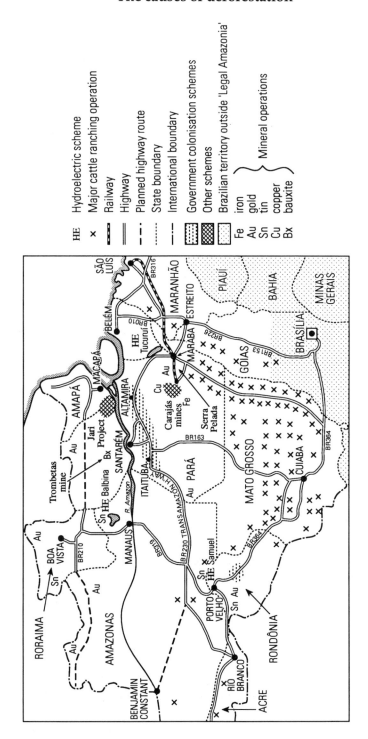

Figure 2.1 *Brazilian Amazonia, showing main highways, ranching and resource exploitation locations*

Fish farming

Considerable areas of mangrove forests in coastal and estuarine areas have been converted to rice paddies but another cause of deforestation in these areas is the introduction of shallow ponds for fish farming. Some 89 per cent of the coastline of the Guayas province of Ecuador, for example, is now occupied by shrimp farms, covering an estimated 100,000 ha in 1986. Shrimps are now Ecuador's second major export after petroleum (Barnett, 1989).

Government resettlement schemes

Governments are intimately involved in promoting the expansion of permanent agriculture. One form of intervention is the official resettlement scheme. Some countries, such as Indonesia and Brazil, have long faced major problems because of high population concentrations in particular areas, so government programmes were introduced to resettle people to less crowded regions, most of which are covered by tropical rain forest. This kind of planned deforestation should be preferable to other forms of deforestation, since ideally land should be selected by taking into account its suitability for continuous cultivation. But this does not necessarily happen in practice so many of the expected benefits are not obtained.

The Indonesian Transmigration Programme

Indonesia has the fourth highest population of any country in the world. Three fifths of its 184 million people live on just 7 per cent of national land area on the island of Java and the neighbouring islands of Madura and Bali. Java has fertile volcanic soils but even so there are limits to its human carrying capacity and a voluntary resettlement scheme, now called the Transmigration Programme, has been operating since 1905. It expanded greatly from the late 1960s onwards, with 2.5 million people resettled between 1978 and 1983 alone. Another aim was to increase national self-sufficiency in rice. Until the mid-1970s most settlers went to the island of Sumatra but then the scheme was extended to include Kalimantan and Irian Jaya (Figure 2.2) (Indonesian provinces on the islands of Borneo and New Guinea respectively).

Transmigration settlements are large cleared areas, varying in size from hundreds to thousands of hectares, with some as large as 10,000 ha, each containing a village with social, educational and health facilities and adjacent farmland. Every family receives a 3.5 ha plot of land, comprising a house area and land to cultivate dryland crops, paddy rice and cash crops (Arndt, 1983). Most settlements have been located in forested areas and between 1978 and 1983 the total area cleared may have averaged up to 280,000 ha per annum.

The Transmigration Programme has not lived up to expectations. Even at its peak in the late 1970s it took five years to resettle the equivalent of three quarters of Indonesia's annual population increment and its contribution to national rice production has been minimal. Agricultural productivity has been generally low as sites were chosen more to meet resettlement targets than to satisfy land suitability criteria. Forest and land were often cleared by bulldozers, which damaged the soil, led to erosion, and removed a lot of the inherent fertility even before the first crop was planted. Rice production, involving huge areas of the same varieties, has been frequently hit by insect infestation. This all led to widespread popular dissatisfaction and up to a fifth of all settlers left transmigration schemes either to return home or clear more forest elsewhere. In the second half of the 1980s the government had to cut back the scale of the programme when income from oil and other commodity exports fell and it could no longer meet the high costs involved (Madeley, 1987).

Resettlement schemes in Brazil
Brazil's population is also highly concentrated: the southeast and northeast regions contain 72 per cent of the total population on just 29 per cent of national land area. In 1970, to coincide with construction of the TransAmazonian Highway, the Brazilian government introduced the National Integration Programme (PIN) to resettle people from overcrowded areas. The first planned settlements, grouped in three Integrated Colonization Projects (PICs), were along the TransAmazonian Highway. A seemingly ideal hierarchy of villages, small towns and large towns was to be built. Each village would have 50 houses, a school, medical and social centres, a store, government offices and a church. Settlers were given low-price houses and 100 ha farm plots (the cost repayable over 20 years), removal expenses and up to eight months salary.

However, PIN was a failure. Compared with the resettlement target of 100,000 families in five years just 7,900 families were moved in eight years, and only a third came from the north-east region. Many of those who did come (up to a third of all settlers in the PICs in Altamira) later left as a result of disease, poor crop yields, lack of promised roads and other facilities and inadequate agricultural support (Smith, 1981; Mougeot, 1985; Moran, 1985). Five other PICs along BR 364 in Rondônia were more successful because soil fertility there was higher. After this experience, in 1974 the government changed the focus of settlement activity to BR 364, where a new

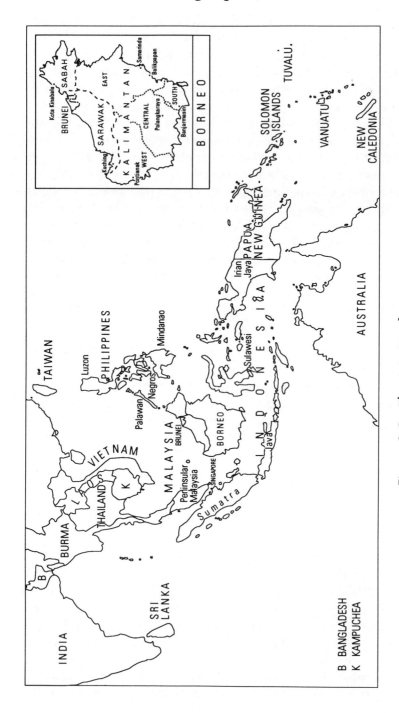

Figure 2.2 The Asia-Pacific region.
Inset shows the island of Borneo divided into the territories of Brunei, Malaysia (Sabah and Sarawak) and Indonesia (Kalimantan)

scheme of Directed Settlement Projects (PADs) merely granted titles to lands. The expensive village infrastructure and financial inducements of the PICs were thought unnecessary given the high rate of spontaneous settlement in Rondônia.

At the same time the government shifted the main emphasis of its Amazonian development programme from smallholdings to large-scale estates, and a new wave of cattle ranching and mining projects began (see below). Further planned settlement projects were established later in Rondônia as part of the Northwest Brazil Integrated Development Scheme (POLONOROESTE) which began in 1979. The continued high rate of migration to Rondônia – averaging 10,000 families every year – also forced the government as part of this scheme to grant spontaneous settlers legal title to lands they had occupied (Fearnside, 1984, 1987). The PIC and PAD resettlement schemes resulted in over 1.6 million ha of forest clearance, or about 11 per cent of all Amazonian forest cleared by 1983 (Browder, 1988).

Cattle ranching

The expansion of cattle ranching in Latin America, and particularly in Brazilian Amazonia, has received a great deal of publicity as a major cause of deforestation, and justifiably so. Between 1966 and 1983 it accounted for about three fifths of Latin America's annual deforestation rate and two thirds of the rate for Brazilian Amazonia alone. Ranching is also widespread in Central America, where it accounted for a third of all deforestation in Costa Rica, Guatemala, Honduras and Nicaragua between 1960 and 1980 (Myers, 1985a). Ranching in that region spread mainly because of a rise in demand for beef imports by the USA, which obtains three quarters of its imports from Central America. Brazil, on the other hand, exports mostly to Western Europe rather than the USA, and though it is now the world's fourth largest beef exporter (Dullforce, 1990) the majority of its beef is produced outside Amazonia.

Raising beef production was one reason why the Brazilian government promoted cattle ranches in Amazonia in the 1970s. Another was the need to achieve large-scale settlement in the region to satisfy national development aims and prevent possible invasions of its territory by people from neighbouring countries. The ranches, which were concentrated on Amazonia's eastern and southern fringes, averaged 23,600 ha in area but some extended to hundreds of thousands of hectares. Ranching in Amazonia grew in two phases, the first between 1966 and 1972, the second from 1974 onwards. Companies participating in schemes organized by the Superintendency for the Development of the Amazon (SUDAM) bought land at low prices and received generous tax incentives. Other ranches were established with government-subsidised loans at interest rates lower than the rate of inflation.

The high hopes that ranching would be sustainable and productive in Amazonia (Falesi, 1974) were not realized. Pasture yields dropped owing to poor (and declining) soil fertility and weed proliferation. This, and the affliction of cattle by various diseases, led to low beef yields. Most ranchers did not invest in the fertilizers needed to maintain pasture yields, choosing instead first to burn pastures regularly to add nutrients to the soil and then to clear and burn more forest when this became insufficient to maintain yields (Homma et al, 1978; Jordan, 1985). Ranchers were supposed to retain forest cover on at least half of their properties but many got around this restriction by clearing half of the forest and then selling the remaining, forested, portion to other ranchers who could then clear half of the remainder themselves, and so on.

With average productivity just a quarter of that on US ranches and overall beef production only 9–16 per cent of projected levels, Amazon cattle ranching has not been profitable overall (Myers, 1985a; Hecht, 1980, 1983). But owing to large government subsidies (amounting to some \$731 million in all between 1966 and 1983) and a rise in land values, many individual enterprises were profitable, and this made the low productivity of academic interest only. Without the subsidies the average ranch would have shown a net loss equivalent to half the money invested in it (Browder, 1985, 1988; Mahar, 1989). Government incentives for new ranches ended in 1979 but existing ranches still received subsidies for years afterwards.

Tree crop and other plantations

Cattle ranching is not an important cause of deforestation in Southeast Asia and Africa. There the equivalent form of large agricultural estate is the tree crop plantation, but tree crops are grown on smallholdings too (Williams, 1975). Among the leading tree crops are rubber, oil palm, cocoa, coffee and coconut. Other plantation crops are sugar – an important cause of tropical deforestation since Europeans first colonized the Caribbean in the sixteenth century – and timber. Tree crop plantations are a legacy from colonial times but in most tropical countries after independence landowners chose, with government backing, to continue them and national plantation areas have expanded over the years. Of the ten leading primary commodities exported by developing countries four are tree crop products: coffee, rubber, cocoa, and vegetable oils such as palm oil and coconut oil (aggregated with non-tree crops). The high capital investment needed usually ensures that plantation owners (which include both private companies and government bodies) select good sites and manage plantations well, so they represent sustainable uses of deforested land, but the large areas of monocultures are susceptible to attack by pests and diseases.

Rubber is made from latex tapped from the rubber tree *Hevea brasiliensis*, which grows naturally in the southern Amazonian territories of Brazil, Peru and Bolivia. A limited amount of tapping of wild trees still takes place there but 93 per cent of all world production comes from plantations in Southeast Asia, where three countries, Malaysia, Indonesia and Thailand, account for 75 per cent of all production. Rising demand for vehicle tyres led to a 'rubber boom' in Amazonia between 1870 and 1912 but Latin America's monopoly ended after seeds were shipped to Asia and used to establish plantations in Peninsular Malaysia and Java that could produce rubber for a quarter of the cost of that from wild trees. The cost differential grew even wider after selective breeding raised yields. Plantations were attempted in Brazilian Amazonia, most notably by Henry Ford in the 1920s, but failed when the trees were attacked by South American leaf blight. Rubber is an ideal crop for small farms as well as large plantations and up to 70 per cent of Indonesian rubber production is grown on smallholdings. Farmers in remote areas with intermittent access to markets like it because once the latex has been cured into rubber sheets these can be stored for some time without spoiling.

In the late 1980s rubber tappers in Brazilian Amazonia gained considerable publicity when they pleaded to their government to establish formal reserves so they could maintain their livelihoods in the face of the spread of cattle ranching and other land uses. In the end they realized their aims but at the cost of the lives of some of their leaders, including Chico Mendes who was murdered by cattle ranchers in 1989. This is but one example of how changes in land use in Amazonia over the last thirty years have been accompanied by often violent conflicts between different groups of people, with the weakest, such as indigenous peoples and rubber tappers, generally losing ground to more empowered ones, such as cattle ranchers and large corporations. The government eventually stepped in to defend the rubber tappers, but since half of Brazil's rubber consumption is imported and local production was until recently heavily subsidized the future of Brazil's rubber industry, even with these latest initiatives, is not very rosy and will ultimately depend upon the breeding of disease-resistant plantation hybrids rather than the continuation of forest tapping.

Early supplies of palm oil, used to make margarine, cooking fats, soap and detergents, came from wild oil palm trees in the rain forests of West Africa, but these were also later overtaken by plantations, first in Zaire and later Southeast Asia. Today Malaysia accounts for two thirds of world production and established 60,000 ha of new plantations in 1989 alone. The share of Indonesia, the second largest producer, is rising fast (Hoon, 1989c). Large monoculture plantations linked to nearby processing plants are needed for efficient production: a typical estate in the Malaysian state of Sabah has five 1,000–2000 ha plantation units plus workers' quarters and smallholdings.

Of the other plantation crops, coconut palm is grown on small and large plantations in Asia, which account for 85 per cent of all production. Coffee cultivation is more widespread, with Brazil and Colombia the leading coffee producers. The centre of the Brazilian industry is the state of Minas Gerais, outside Amazonia, but coffee is popular on resettlement schemes in Rondônia (CEPA-RO, 1980). The cocoa tree grows wild in Brazilian Amazonia, but there are now plantations all over the world, including those on smallholdings in resettlement schemes in Rondônia and along the TransAmazonian Highway (Fearnside, 1985b). Sugar cane plantations were historically a major cause of deforestation in Brazil, the Philippines and Cuba, especially, but Indonesia and Colombia now have similar plantation areas to those in the Philippines, and Thailand has half as much again.

Timber is another key plantation crop (Evans, 1982). In some areas plantations have been established on land that had been deforested for some time, but in Papua New Guinea, Sabah (which had 13,000 ha of plantations by the early 1980s) and elsewhere tropical rain forest had to be specially cleared so it could be 'reforested'. Soil fertility should be greater on such sites than on degraded lands and the chances for sustainable production better. About 0.7 million ha of new timber plantations were established annually in the humid tropics in the late 1970s but how much of this required fresh deforestation is difficult to say. Only one fifth of all present plantations produce high quality hardwood equivalent to that from tropical rain forests. The majority produce pulpwood, fuelwood and softwood timber instead (Grainger, 1986). Timber plantations are discussed in more detail in Chapters 3 and 7.

Other types of deforestation

Agricultural expansion is not the only reason for deforestation. Mining, hydro-electric power schemes and illegal cultivation of narcotic plants[2] also contribute to the problem. In more developed countries urban expansion plays a part too.

Mining

Deforestation occurs when the rich mineral deposits of the humid tropics are exploited, though the extent of deforestation depends on whether opencast (strip) or underground (shaft) mining is used. The larger opencast mines cover thousands of hectares, wastes from mining and mineral processing pollute rivers, and there is concern too about the encroachment of mining activities on to the territories of indigenous peoples.

Tin mining in the humid tropics has a long history. Opencast methods have been most popular but shaft mining has been used in Bolivia and low-cost offshore dredging is now the main practice in Indonesia (Bolderson, 1990, 1991). Malaysia, Indonesia, Bolivia and Thailand used to control 80 per cent of world production but in the 1980s Brazil and China emerged as major new producers. Production in Brazil rose steadily as it was still profitable when prices fell in the 1980s due to reduced demand and surplus production. Vast new reserves were discovered and exploited in Amazonia, first at Pitinga, 250 km north of Manaus (then the world's largest tin mine), and later at Bom Futuro in Rondônia, which was even larger (Figure 2.1). A takeover of Bom Futuro by 20,000 freelance miners in 1988 forced the Brazilian government to ask the owners of Pitinga (who also held the rights to Bom Futuro) to cut back production to prevent further chaos in the world tin market. Brazil became the world's leading tin producer but Bom Futuro was a major headache for the government, owing to increasing violence, suspicions that miners were involved in drug trafficking, and heavy pollution of local rivers by sludge left after tin ore had been removed. An estimated 5,000 ha of forest were cleared there. The mine was closed by government order in 1991 and the miners evicted. The decision was later rescinded but most miners had left by then to look for new tin sources in Roraima in northern Amazonia (Barham, 1989a, 1989c, 1990a, 1990b, 1990c, 1990d; Griffith, 1991a, 1991b, 1991c; Knight, 1987a, 1987b).

Brazil is now the world's fifth largest gold producer, most of its annual output of about 100 tonnes coming from small independent operators (Barham, 1989b). At the Serra Pelada mine in the Carajás region of Pará state miners pan gold from river silts or excavate small claims on the slopes of huge pits. Other mines have been opened nearby, and in the states of Goiás, Roraima and Rondônia (Salati et al, 1987; Ellis, 1988). In Rondônia, miners on barges moored in the River Madeira pump up mud from the river bottom using suction pipes. The dumping of mercury used in processing gold has led to severe river pollution problems (Vanvolsem, 1990). Gold is also mined in neighbouring countries. Papua New Guinea and the Philippines are the leading Southeast Asian producers but Indonesia's output should rise quickly following finds in East Kalimantan and Irian Jaya in 1986. Almost all Papua New Guinea's gold production comes from large open-cast copper mines at OK Tedi and Panguna (on the island of Bougainville – see below) but a new underground gold mine opened in 1990 at Porgera, close to OK Tedi, and other promising reserves have been found elsewhere too.

Copper mining has been very prone to disruption over the last thirty years. Zaire and Peru are the two largest producers in the humid tropics but their combined output is less than half that of Chile, the world's top producer. Peru's mines were the target of frequent terrorist attacks in the late 1980s and mining in Zaire was disrupted in 1991 as the country sank into political turmoil. A large open-cast mine at Panguna on the island of Bougainville, which opened in 1972, required the clearance of over 9,000 ha of tropical rain forest and resulted in a huge crater. After a prolonged campaign of violence and sabotage by militants calling for the closure of the mine and secession of Bougainville from Papua New Guinea mining operations were suspended indefinitely in May 1989 and Australian employees left the following year. A peace treaty was signed between the government and rebels in 1991 but the future of the mine is still uncertain (Sherwell, 1990a, 1990b; Brown, 1991).

The West African country of Guinea is the world's second largest exporter of aluminium ore (bauxite) after Australia. Two thirds of its output comes from a huge opencast mine at Sangaredi in the northwest of the country, the bauxite being exported to Europe and North America for smelting into aluminium as there is no cheap local electrical power. Indonesia not only has bauxite reserves but also has two hydroelectric power stations on the island of Sumatra to power a local smelter, the largest in Southeast Asia. Latin American producers include Venezuela, Surinam, Guyana and Brazil, which has a large opencast bauxite mine in north-east Amazonia 80 km from the confluence of the River Trombetas with the River Amazon and 483 km east of Manaus. It covers 457,000 ha and forest there is cleared at a rate of 618 ha per annum. A reforestation programme restores the land with 90 native and 12 exotic species when mining has finished, using as much of the original soil and rock as possible. Bauxite from the mine is shipped to an aluminium smelter near Belém, at the mouth of the River Amazon, which uses electricity from the Tucuruí hydroelectric scheme in eastern Amazonia. Brazil's aluminium production almost tripled between 1982 and 1987 (Blackburn, 1987).

The Carajás mountains in the north-east of Brazilian Amazonia contain one of the world's largest known deposits of iron ore and a huge mine was opened there in 1983 to exploit them. The mine is part of a wider regional development programme, the Programa Grande Carajás, which also includes the Tucuruí hydroelectric scheme, the Belém aluminium smelter and various agricultural projects (Eden, 1990; Hall, 1991). Ore is shipped from the mine to the port of São Luís along an 876 km railway line and Brazil now exports more iron ore than any other developing country. The company owning the mine, Companhia Vale do Rio Doce (CVRD), has taken care to minimize its environmental impacts (Mahar, 1989) but there is great concern that rising local populations will cause deforestation, that tropical rain forest through which the railway line passes will be affected by dust and soil erosion, and, even worse, that 21 small iron ore smelters to be sited along the railway line will use wood from the forest, rather than from plantations, as their basic energy feedstock. If so then 82,000 ha of rain forest could be cleared each year for this purpose alone (Fearnside, 1988, 1989a). Six of the smelters were already in operation by 1990 (Lamb, 1990).

Opencast coal mines are an increasingly important cause of deforestation in Latin America and Southeast Asia. The largest steam coal mine in the world is at El Cerrejon in Colombia. It and other nearby mines produce coal with a low sulphur content, attractive to industrialized countries since it should help to alleviate acid precipitation and other environmental impacts caused by the emission of sulphur dioxide and other gases from coal-fired power stations (McCloskey, 1990). Coal mining is also expanding in Indonesia, which has proven reserves in Sumatra, East Kalimantan, South Kalimantan and Irian Jaya. Coal from Kalimantan also has a low sulphur content and three opencast mines were being developed in the early 1990s, as were ocean terminals to ship the coal all over the world (McCloskey, 1990). The eventual impacts of these mines on the valuable rain forests of East Kalimantan is uncertain.

A number of countries in the humid tropics, including Cameroon, Congo, Gabon, Nigeria, Indonesia, Malaysia, Ecuador and Venezuela, are major exporters of petroleum and/or natural gas.

Some are leading members of the Organization of Petroleum Exporting Countries (OPEC). The amount of deforestation associated with each oil well is less than that for opencast mines, but the building of roads, pipelines, and testing and well sites for oil exploration also has a significant impact on forests. The oil industry has caused considerable deforestation in the north of Ecuadorian Amazonia over the last 20 years, encroached on to the lands of the Waorani people and the Cuyabena National Park, and led to oil spills and leakages that polluted rivers and lakes and harmed wildlife. The recent southward shift of oil exploration aroused political protests at national and local levels and troops had to be called in to protect oil installations (Kendall, 1990). A lack of domestic petroleum reserves meant that Brazil for a long time had to depend on costly oil imports but its own industry developed rapidly in the 1970s when offshore reserves were discovered. In 1986 and 1987 reserves were also found on the River Urucu in Amazonas state (Charters, 1987).

Hydroelectric schemes

Hydroelectric power is a major renewable energy resource but to exploit it in the humid tropics requires forest clearance or submersion in order to build dams and reservoirs. The leap in oil prices in 1973 prompted the Brazilian government to develop the vast untapped hydroelectric power of Amazonia, estimated at 100 gigawatts (GW), or 100,000 megawatts (MW). (Installed hydroelectric capacity worldwide in 1984 was 542 GW (L'vovich and White, 1990)). By 1988 four hydroelectric projects had been completed, three were being constructed (Barrow, 1988) and 22 GW of installed capacity was planned by the year 2000 (Holtz, 1989). However, this expansion has aroused opposition at home and abroad. In the late 1980s protests by the Caiapó Indians, ecologists and environmentalists made the World Bank drop plans to fund an 11,000 MW dam on Caiapó land on the River Xingu at Babquara in Pará state (Anon, 1988b; Dawnay, 1989a).

The largest dam in Brazilian Amazonia so far is at Tucuruí on the Tocantins River, 300 km south of the port of Belém (Figure 2.1). Completed in 1985 at a cost of over $4 billion it should ultimately generate 4000 MW, eclipsing the earlier Paredão and Curuá-Una projects (40 MW and 20 MW respectively). It supplies power to various industrial customers, including the aluminium smelters near Belém. The 2000 sq km reservoir was much larger than those of the two earlier schemes (approximately 100 sq km each). The site was to have been totally deforested before flooding since hydrogen sulphide given off by decomposing vegetation is a human health risk and in previous projects made dam water so acidic that turbine parts corroded, requiring costly repairs (Bunyard, 1987; Caufield, 1984). But only part of the forest was cleared, and the remainder was treated with defoliant instead. The contractor involved went bankrupt, claiming there was too little commercial timber in the forest to make clearing profitable. By way of compensation he was later allowed to log over 90,000 ha in two nearby Indian reserves (Pereira, 1982; Barham and Caufield, 1984).

Cultivation of illegal narcotics

Deforestation is also caused when narcotic plants such as coca (the source of the illegal drug cocaine) and opium (the source of heroin) are grown in small tropical rain forest clearings in Southeast Asia and South America. Often the practice is similar to that of shifting cultivation, for when the plots are discovered by drug enforcement officers the farmers try to move elsewhere before being apprehended. The scale of deforestation caused in this way is quite significant. Peru had an estimated 200,000 ha of coca plantations in 1987 (Tyler and Kendall, 1987) and is the leading producer of coca, accounting for 60 per cent of world output, followed by Bolivia, Colombia and Ecuador (Graham, 1988). The dumping of chemicals used in processing coca pollutes local rivers. Opium grown in remote mountainous and hilly areas in Thailand, Laos and Burma accounts for 70 per cent of world production. Half of this comes from Burma.

Urban expansion

As countries get more developed urban-industrial expansion becomes a significant cause of deforestation. In Thailand, for example, forest is being cleared for the construction of second houses for urban dwellers, golf courses and other tourist developments. Urban expansion also takes place there at the expense of agricultural land, but this can lead to more deforestation as new (and often less fertile) farmland is cleared from forest in other parts of the country.

Land use change and deforestation

Contrary to popular perceptions, a wide range of land uses can replace forest after deforestation. The dominant land uses differ from region to region: for example, cattle ranching is a major cause of deforestation in Latin America but not in Africa. Both shifting and permanent agriculture are expanding at the expense of forests but forest is also cleared for other land uses, such as mining, hydro-electric power projects, the cultivation of narcotic plants and urban expansion. Each land use differs in the scale of forest clearance required, the period of time over which forest is cleared, and its own sustainability at site level – which in turn affects future trends in deforestation. Changes in land use are not wrong in themselves, but if the new land uses prove unsustainable and more deforestation is required then the justification for the original deforestation is devalued. This chapter focused on describing how deforestation takes place, but all land use change is a response to demand to use that land to produce goods and services other than those accruing from the natural forest. This, in turn, results from socio-economic factors and government policies whose influence is discussed in Chapter 4.

3

Logging and the tropical hardwood trade

Logging, the second major human impact on the tropical rain forests, is essential if they are to be managed as renewable sources of wood, for mature trees need to be felled and removed from the forest so that young trees can grow up to replace them. Not all tropical rain forest has to be managed in this way of course but timber production does give an incentive to protect large areas from being cleared. The tropical rain forests contain vast reserves of high quality timbers, some of which have been commercially exploited for hundreds of years, but exploitation has only been widespread and intensive since 1945 in response to the rapid rise in overseas and domestic demand for tropical hardwood. Although selective logging degrades the forest rather than causing deforestation as such, there is concern that poor logging practices result in too much forest damage and, together with the limitations of present selective logging systems, mean that forest management is currently not as sustainable as we would like. After distinguishing between deforestation and logging, this chapter gives a brief introduction to the tropical hardwood trade, describes selective logging systems and how they work in practice, and ends by discussing the sustainability of current logging practices and the prospects for future improvements.

Deforestation and logging

Logging in tropical rain forests differs from deforestation since the almost universal practice is not clearfelling, as in temperate forests where whole stands are cleared to harvest timber, but selective logging. Only a few of the thousands of tree species in tropical rain forests are currently in commercial demand so typically only between two and ten trees of commercial species are felled and removed per hectare out of a total of over 350 trees.[1] Some forest is cleared to construct tracks, roads, landings, etc., but apart from this when the loggers leave the forest still remains. Deforestation has not occurred, therefore, but the forest has been degraded in structure and species composition (see Chapter 1). The extent of degradation depends on the skill of the loggers, but the forest usually retains its capacity to regenerate as a renewable resource so that more timber can be extracted at the next harvest in 30 to 40 years time.

Logging and deforestation are therefore best treated as two distinct phenomena, which together give a good approximation to the total human impact on the tropical rain forests. There are other reasons for treating them separately. While logging is not a direct cause of deforestation it can be an indirect cause, for after the loggers have departed other people may enter the forest along logging roads and clear it for agriculture. The nature of the demand for land for agriculture is also very different from that for timber. Different but complementary strategies are therefore needed in the agriculture and forest sectors to control deforestation. Improving forest management alone is not sufficient. Finally, deforestation can be easily monitored on a global scale by satellite remote sensing techniques but monitoring logging in this way is more difficult.

What is tropical hardwood?

Most trees in tropical rain forests are broadleaved, like the oaks in temperate forests, rather than coniferous (or needle-leaved), like the pines. So while there are some important coniferous genera, such as *Agathis* and *Araucaria*, all but a few of the commercially marketable timbers extracted from tropical rain forests are hardwoods, not softwoods. 'Tropical hardwood' refers here, for convenience, to all tropical hardwoods as an aggregated commodity.

Half of all wood harvested in the world is *industrial wood*, i.e. the logs are later converted into products such as sawnwood, plywood and veneers (thin strips of wood 'peeled' from logs) (Table 3.1). Most of the trees logged in tropical rain forests are converted into these products for use in construction and furniture manufacture. The other half of world wood production is used as *fuelwood* or charcoal. Wood is still a major fuel in developing countries, where it accounts for 80 per cent of all wood harvested (compared with 20 per cent in developed countries). Fuelwood is gathered in tropical rain forests by local people but only accounts for a small share of total removals and is not a significant cause of deforestation there. The situation is different in forest-poor dry and montane tropical regions (Grainger, 1990a), many of which suffer from fuelwood shortages. Those seldom occur in the humid tropics except in parts of a few countries, such as the Philippines and Thailand, where population density is high and remaining forest cover low.

Table 3.1   *World wood removals in 1987 (million cubic metres)*

	Total	Developed countries	Developing countries
All roundwood	3,380	1,526	1,854
Industrial roundwood	1,644	1,249	395
Fuelwood and charcoal	1,736	277	1,459

Source: FAO (1989)

Tropical hardwoods vary greatly in density and other physical properties, and in their end uses. The most popular tropical hardwoods are now the low and medium density utility woods, so called as they have a wider range of uses than decorative woods such as teak and mahogany, that dominated exports before the Second World War. Woods converted into plywood and veneers include okoume, obeche, limba and makore from Africa, and lauan, meranti and seraya from Asia (Table 3.2). Together with abura, iroko, kokrodua and niangon from Africa and ramin and keruing from Asia they are also made into sawnwood for use in construction. The highest quality decorative veneers and sawnwood come from mahogany, sapele and teak, the heavier kokrodua and utile, and selected logs of other species. Heavier and very durable woods used for key structural purposes in the construction industry include greenheart, ekki and keruing. All the above timbers are high-grade hardwoods, in contrast to lesser quality woods used only for fuelwood or pulpwood.

The post-war development of the tropical hardwood trade

Rising demand for imports

Extensive logging of tropical rain forests is a post-Second World War phenomenon. Though it has been catalysed by the growth in demand for imports by developed countries, half of all tropical hardwood removals is still consumed in the producing countries themselves[2]. Tropical hardwood exports rose fourteen-fold from 4.6 to 61.2 million cubic metres rwe (roundwood equivalent volume)[3] per annum between 1950 and 1980 (Figure 3.1), when they accounted for 45 per cent of all removals, estimated at 135 million cubic metres per annum. Before the Second World War world trade in tropical hardwood was limited to European (and latterly US) imports of a

Table 3.2. *Some leading tropical hardwoods*

Botanical name	Common name	Use class*
Africa		
Aucomea klaineana	Okoume/Gaboon	LU
Chlorophora excelsa	Iroko	LU
Entandrophragma cylindricum	Sapele	D
Entandrophragma utile	Sipo/Utile	D
Khaya spp.	African mahogany/Acajou	D
Lophira alata	Ekki	C
Lovoa trichiliodes	African walnut	D
Mitragyna ciliata	Abura	LU
Pericopsis elata	Kokrodua/Afrormosia	D
Tarrietia utilis	Niangon	LU
Terminalia superba	Limba	LU
Tieghemella heckelii	Makore	HU
Triphlochiton scleroxylon	Obeche	LU
Asia-Pacific		
Dipterocarpus spp.	Apitong[P]/Keruing[MSI]	C
Gonystylus spp.	Ramin	HU
Parashorea malaanon	White lauan[P]/White seraya[S]/	LU
Pentacme contorta.	White lauan[P]	LU
Shorea negrosensis	Red lauan[P]	HU
Shorea spp.	Light red meranti[MI]/Light red seraya[S]	LU
Shorea spp.	Dark red meranti[M]/Dark red seraya[S]	HU
Tectona grandis	Teak	D
South America		
Balfourodendron riedelianum	Pau marfim	HU
Calophyllum brasiliense	Jacareuba	LU
Carapa guianensis	Andiroba/Crabwood	LU
Cedrela odorata	Cedro-Vermelho/Red cedar	LU
Cordia goeldiana	Freijó	D
Ocotea rodiaei	Greenheart	C
Swietenia macrophylla	Mahogany	D
Virola spp.	Virola	LU

NB For the Asian species common names are given separately for Indonesia (I), Malaysia except for Sabah (M), Sabah (S) and the Philippines (P). The common names for most South American species are from Brazil.
* *Use classes:* D = Decorative; LU = Light Utility (low to medium density); HU = Heavier Utility (upper to high density); C = Constructional (high, very high or exceptionally high density).
Source: Pitt (1981), with use classes based on Hansom (1981).

Logging and the tropical hardwood trade

few decorative woods such as mahogany (*Swietenia* spp.) from Central America, teak from South-east Asia and ebony from Africa. But in the aftermath of the war tropical hardwood was substituted for temperate hardwood in the European market when US dollars to purchase American timber were scarce. Tropical hardwood established a market niche as a utility wood in the construction and furniture industries, and when demand for such wood grew in the 1950s and 1960s, first in Europe and then in Japan, tropical hardwood was bought because it was available in attractive qualities and large quantities at competitive prices.

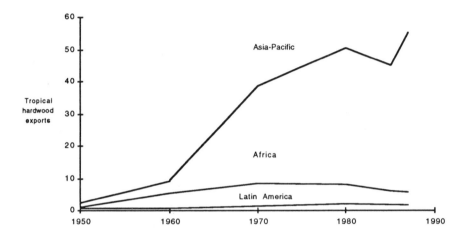

Source: Grainger (1986), based on FAO data

Figure 3.1 *Trends in tropical hardwood exports 1950–87 (million cubic metres per annum)*

Despite Europe's key early role Japan was the main driving force for the surge in exports. In the late 1950s it began a period of massive economic reconstruction and tropical hardwood imports grew by a factor of 23 between 1950 and 1960 alone, taking Japan's share of all tropical hardwood imports to a third, compared with only 5 per cent just after the war. By 1970 its imports had quadrupled to 20 million cubic metres rwe and have stayed near that level ever since (Table 3.3). The present European Community countries took 38 per cent of all imports in 1950 but while their imports increased sixfold between 1950 and 1980 market share dropped to 17 per cent. US imports quadrupled between 1950 and 1970 but only to about 3 million cubic metres rwe and its share of all imports fell from 16 per cent to about 6 per cent, a share it has retained since[4].

Table 3.3. *Trends in tropical hardwood imports 1950–87 (million cubic metres roundwood equivalent volume)*

	1950	%	1960	%	1970	%	1980	%	1987	%
Total imports	4.364		15.759		48.832		61.226		62.822	
Logs	2.445	56	10.670	68	36.720	75	38.475	63	28.690	46
Sawnwood	1.792	41	3.744	24	6.722	14	13.857	22	17.155	27
Plywood/veneers	0.127	3	1.345	8	5.390	11	8.894	15	16.977	27
Japan	0.209	5	4.744	30	20.652	42	19.300	32	19.208	31
Logs	0.098	47	4.729	100	19.875	96	18.365	95	14.235	31
Sawnwood	0.111	53	0.015	0	0.428	2	0.775	4	1.578	8
Plywood/veneers	0.000	0	0.000	0	0.439	2	0.160	1	3.394	18
EC	1.663	38	5.294	34	8.549	18	10.277	17	8.171	13
Logs	0.840	51	3.644	69	5.442	64	4.298	42	1.557	19
Sawnwood	0.785	47	1.430	27	2.344	27	4.694	45	4.696	57
Plywood/veneers	0.038	2	0.220	4	0.763	9	1.285	13	1.918	23
USA	0.680	16	1.989	13	3.144	6	2.681	4	3.461	6
Logs	0.430	63	0.326	16	0.173	5	–	–	0.000	0
Sawnwood	0.239	35	0.698	35	0.689	22	0.597	22	0.525	15
Plywood/veneers	0.011	2	0.965	49	2.282	73	2.084	78	2.936	85
Persian Gulf	na	–	na	–	na	–	0.373	1	0.728	1
Other Asia	na	–	na	–	6.658	13	13.571	22	12.181	19
Japan/EC/USA	2.552	59	12.027	76	32.345	65	32.258	53	30.839	49
TOTAL	2.552	59	12.027	76	39.003	78	46.202	75	43.748	70

NB 1: % in national totals refer to % total exports. % in products refer to % national totals. Persian Gulf imports refer only to plywood; Other Asia imports to logs. See text for definition of Other Asian Processing Nations. EC = European Community

2: Conversion factors from product volume to roundwood equivalent volume: logs 1.00, sawnwood 1.71, plywood and veneers 1.82.

3: Totals are all exports of a product. Percentages refer to national percentages of all imports/exports of a product.

4: Owing to poor data, plywood/veneers imports for 1950 and 1960 refer only to plywood imports; EC log imports for 1950 and 1960 exclude Greece.

Source: Adapted from Grainger (1986)

Almost three quarters of the European Community's tropical hardwood imports are now taken by just four countries. The Netherlands and the UK were the leaders in 1987, together taking 40 per cent of all imports (Table 3.4). France and Italy were next with 32 per cent between them. This is a recent change from the earlier pattern when the trade was dominated by France, Italy and the former West Germany. They took 60 per cent of all imports in 1980 but a large proportion of these were logs. As log availability tightened and plywood supplies increased in the 1980s these three countries lost ground as the leading plywood importers, the UK and the Netherlands, doubled their imports of that product.

Table 3.4. *Leading European Community importers of tropical hardwood in 1987 (thousand cubic metres roundwood equivalent volume)*

	Volume	% Total
Netherlands	1,802	22
UK	1,443	18
France	1,406	17
Italy	1,230	15
Belgium/Luxembourg	961	12
Germany	615	8
Spain	483	6
Portugal	167	2
Denmark	64	1

A major change took place in the tropical hardwood trade in the 1970s with a rise in imports by a group of rapidly developing Asian countries – Hong Kong, South Korea, Singapore and Taiwan – referred to here as the Other Asian Processing Nations. They built lots of mills to convert imported tropical hardwood logs into sawnwood, plywood and finished products such as furniture. Much of the output was exported but substantial domestic markets developed too. By 1980 the Other Asian Processing Nations took 22 per cent of all tropical hardwood exports and together with Japan, the EEC, and the USA they accounted for more than three quarters of all exports. Their exports later declined, due mainly to the banning of log exports from Indonesia. Two other growth markets were the Persian Gulf countries and the People's Republic of China.

Tropical hardwood's world role

Tropical hardwood plays a leading role in the world forest products trade. The wide distribution of forests on our planet means that most countries supply a good proportion of their own wood requirements so only 25–33 per cent of all wood removals is traded internationally. Trade is still necessary, however, to exchange wood products available in some countries but not in others (such as tropical hardwood) and to supplement domestic supplies.

Europe, Japan and the USA take almost three quarters of all forest products imports. The European Community's share is the largest (43 per cent in 1987) as its remaining forest cover is insufficient to provide more than half its wood needs. The USA is a major forest product exporter but still took 16 per cent of all imports in 1987 owing to high demand for certain wood products. Japan's share was 12 per cent, for though it has a high percentage of national forest cover importing logs is seen as more cost effective than exploiting domestic forest resources (Grainger, 1993; FAO, 1989).

In 1987 tropical hardwood accounted for 25 per cent of all log exports in the world, 88 per cent of all high-grade hardwood log exports, 11 per cent of all sawnwood exports, 65 per cent of all hardwood sawnwood exports, 39 per cent of all wood panel exports (i.e. including particle board and other products as well as plywood and veneers) and 70 per cent of all plywood and veneer exports. While tropical hardwood exports (in roundwood equivalent volume) only accounted for about 4 per cent of world industrial wood removals the share of high-grade hardwood removals was 23 per cent (Grainger, 1986, 1993).

Major trade flows

Rising demand for tropical hardwood imports after 1950 was not met by all tropical regions and countries uniformly (Table 3.5)[5]. Although the leading exporting countries in each region changed over time there were four major regional trade flows between 1950 and 1980: logs were shipped from Asia-Pacific to Japan and the Other Asian Processing Nations, and from West and Central Africa to Europe; sawnwood and plywood from Asia-Pacific and Other Asian Processing Nations to Europe and North America; and sawnwood from Latin America to North America. After 1980 the pattern started to become more complex, for instance logs began to be exported from Africa to Japan as Asian log supplies became tighter.

Until recently, Japan relied almost totally on imports of logs rather than processed tropical hardwood (sawnwood, plywood or veneers). The USA has imported mainly processed wood for some time, with Europe taking a mixture of logs and products. Logs therefore used to dominate tropical hardwood exports, but their share fell from 63 per cent in 1980 to 46 per cent in 1987, due mainly to a dramatic shift in the 1980s when Indonesia stopped exporting logs to focus on locally produced plywood instead. The role of domestic processing is discussed in Chapter 4.

Table 3.5. *Major tropical hardwood trade flows in 1987*

A. Percentage of all imports by source				
	Japan	Europe	USA	Other Asia
Africa	6	14	1	4
Asia-Pacific	92	69	71	96
Other Asia	2	12	13	0
L. America	0	5	15	0
B. Percentage of all exports by destination				
	Japan	Europe	USA	Other Asia
Africa	21	20	46	14
Asia-Pacific	42	13	1	44
Other Asia	23	59	18	0
L. America	0	44	56	0
C. Percentage of all imports by product and source				
		Japan	Europe	USA	Other Asia
Logs	Africa	8	29	0	8
	Asia-Pacific	92	57	0	92
	L. America	0	0	0	0
Sawnwood	Africa	0	12	2	0
	Asia-Pacific	91	75	46	100
	Other Asia	9	10	5	0
	L. America	0	4	47	0
Plywood/	Africa	0	5	1	0
veneers	Asia-Pacific	93	58	76	100
	Other Asia	7	26	14	0
	L. America	0	11	9	0

NB The Asia-Pacific region does not here include Other Asia (Other Asian Processors) The inclusion of Other Asia as a separate exporting region leads to double counting of wood originally obtained from mainland Asia-Pacific forests, and also of wood products imported from Asia-Pacific and re-exported

Trends in exports

Logging in the tropical rain forests in the post-war era has been concentrated in the Asia-Pacific region. In 1950 it accounted for half of all tropical hardwood exports, with the rest about equally divided between Africa and Latin America. But as demand for tropical hardwood rose it supplied most of the 14-fold rise in exports over the next 30 years, increasing its share of exports to 83 per cent by 1980 (Figure. 3.1, Table 3.6). Africa's share fell consistently, reaching 18 per cent in 1970 and 9 per cent in 1987. Latin America exported very little tropical hardwood after 1950, its share of all exports generally remaining below 5 per cent. Asian dominance was due to a variety of factors, principally low logging costs, owing to the high commercial timber content of its forests, and low transport costs owing to the proximity of the major market of Japan. Large investments were made in processing facilities in the Other Asian Processing Nations especially (Grainger, 1986). At the national level, three quarters of all tropical hardwood exports in 1987 came from just two countries: Malaysia and Indonesia. The Philippines, Ivory Coast and Ghana were far behind with only 2–3 per cent each (Table 3.7).

Table 3.6. *Trends in tropical hardwood exports 1950–87 (million cubic metres roundwood equivalent volume)*

	1950	%	1960	%	1970	%	1980	%	1985	%	1987	%
Africa	1.107	25	5.518	35	8.645	18	8.290	14	6.141	12	5.535	9
Logs	0.850	77	4.315	78	6.833	79	6.547	79	4.217	69	3.594	65
Sawnwood	0.239	22	0.966	18	1.277	15	1.226	15	1.358	22	1.426	26
Plywood/ veneers	0.018	2	0.237	4	0.535	6	0.517	6	0.566	9	0.515	9
Asia-Pacific	2.448	56	9.527	60	38.779	79	50.751	83	45.116	85	55.497	88
Logs	1.229	50	6.020	63	29.525	76	31.803	63	22.440	50	25.062	45
Sawnwood	1.146	47	2.470	26	4.593	12	11.248	22	10.805	24	14.691	26
Plywood/ veneers	0.073	3	1.037	11	4.661	12	7.700	15	11.870	26	15.745	28
Latin America	0.809	19	0.714	5	1.408	3	2.185	3	1.556	3	1.789	3
Logs	0.366	45	0.335	47	0.362	26	0.125	6	0.047	3	0.034	2
Sawnwood	0.407	50	0.308	43	0.852	82	1.383	63	0.925	59	1.038	58
Plywood/ veneers	0.036	5	0.071	10	0.194	19	0.677	31	0.584	38	0.717	40
Total	4.634		15.759		48.832		61.226		52.813		62.822	
Logs	2.445	56	10.670	68	36.720	75	38.475	63	26.704	51	28.690	46
Sawnwood	1.792	41	3.744	24	6.722	14	13.857	22	13.088	25	17.155	27
Plywood/ veneers	0.127	3	1.345	9	5.390	11	8.894	15	13.020	25	16.977	27

NB Percentages in the regional total lines and the aggregated section refer to percentages of total exports. Other percentages refer to percentages of regional totals. Conversion factors for roundwood equivalent volume are given in Table 3.3.
Source: Adapted from Grainger (1986).

Table 3.7. *Tropical hardwood exports by country in 1987 (million cubic metres)*

	Logs		Sawnwood		Plywood/ veneers		All wood	
	pv	%	pv	%	pv	%	rwe	% All
Total	**28.690**		**10.032**		**9.328**		**62.829**	
Africa	**3.594**	**13**	**0.834**	**8**	**0.283**	**3**	**5.541**	**9.0**
Cameroon	0.442	12	0.063	8	0.016	6	0.579	0.92
Central African Republic	0.041	1	0.030	4	0.003	1	0.098	0.16
Congo	0.327	9	0.032	4	0.043	15	0.460	0.73
Gabon	0.121	3	0.001	0	0.051	18	0.215	0.34
Ghana	1.229	34	0.173	21	0.022	8	1.566	2.49
Guinea	0.008	0	0.000	0	0.000	0	0.008	0.01
Ivory Coast	0.617	17	0.460	55	0.098	35	1.586	2.52
Liberia	0.251	7	0.005	1	0.000	0	0.260	0.41
Nigeria	0.060	2	0.001	0	0.000	0	0.062	0.10
Zaire	0.176	5	0.026	3	0.024	8	0.264	0.42
Asia-Pacific	**25.062**	**87**	**8.591**	**86**	**8.651**	**93**	**55.497**	**88.0**
Burma	0.147	1	0.061	1	0.000	0	0.252	0.40
Indonesia	0.003	0	2.766	32	5.542	64	14.792	23.54
Malaysia	22.812	91	3.883	45	1.119	13	31.516	50.16
Papua New Guinea	0.170	1	0.004	0	0.000	0	0.177	0.28
Philippines	0.200	1	0.638	7	0.323	4	1.882	3.00
Thailand	0.000	0	0.083	1	0.083	1	0.293	0.47
Singapore	–	–	0.997	0	0.716	8	3.011	4.79
South Korea	–	–	0.015	12	0.176	2	0.344	0.55
Taiwan	–	–	0.055	1	0.549	6	1.088	1.73
Latin America	**0.034**	**0**	**0.607**	**6**	**0.394**	**4**	**1.791**	**3.0**
Bolivia	0.000	0	0.057	9	0.002	1	0.102	0.16
Brazil	0.010	29	0.460	76	0.291	74	1.328	2.11
Colombia	0.000	0	0.002	0	0.002	1	0.007	0.01
Ecuador	0.000	0	0.027	4	0.015	4	0.074	0.12
French Guiana	0.001	3	0.012	2	0.000	0	0.022	0.03
Guyana	0.018	53	0.009	1	0.000	0	0.033	0.05
Peru	0.000	0	0.002	0	0.001	0	0.005	0.01
Surinam	0.000	0	0.003	0	0.004	1	0.012	0.02
Venezuela	0.000	0	0.000	0	0.000	0	0.000	0.00

NB pv = product volume, rwe = roundwood equivalent volume.
Bold regional percentages are of All Tropics. Other percentages are of bold regional totals, except for the final column, when both regional and national percentages are of all exports.
Source: Adapted from Grainger (1986).

The Philippines was the leading Asian log exporter until the mid-

Controlling tropical deforestation

1960s when its place was taken by Malaysia. But Malaysia had to be content with a joint leadership role after Indonesia expanded its exports in the 1970s and in 1980 the two countries supplied 95 per cent of Asian log exports and 78 per cent of all log exports. In 1987, Asia-Pacific still accounted for 88 per cent of all tropical hardwood exports, but following Indonesia's log export ban half of all exports now came from Malaysia and only a quarter from Indonesia. Malaysia now accounts for the vast majority of the region's log exports, most coming from its two states on the island of Borneo, Sabah and Sarawak, rather than Peninsular Malaysia, which concentrates on exporting sawnwood, plywood and veneers. Most of Indonesia's logging is carried out in its so-called 'outer island' territories on the islands of Sumatra, Borneo (the provinces of East, Central, South and West Kalimantan) and New Guinea (Irian Jaya occupies the western half of the island).

Ivory Coast and Ghana were the leading African exporters in 1987, accounting for half of all exports of logs, plywood and veneers, and two thirds of all sawnwood exports. Cameroon and Congo were next in order of volume exported. Ivory Coast has dominated the African trade over the last 30 years, but its overall exports fell by 60 per cent between 1980 and 1987, the government has banned log exports in a response to severe forest depletion, and it is doubtful if the country can remain a major force in the trade for much longer. Zaire still exports very little tropical hardwood.

Brazil is the leading Latin American exporter, accounting for three quarters of all the region's exports of sawnwood, plywood and veneers in 1987 (only a small quantity of logs is exported from the region). Exports from other countries were minimal, although Bolivia and Ecuador supplied 13 per cent of all sawnwood exports. However, until recently most Brazilian hardwood production came from forests outside Amazonia, owing to the high costs of logging remote forests with a low commercial timber content. But this changed in the 1980s, with Amazonian removals rising from 4.5 million cubic metres in 1975 to 17.4 million cubic metres in 1984 (Browder, 1989).

This, in brief, is the commercial background to the expansion of tropical rain forest logging after the Second World War. It is important to emphasize that tree species are only logged when their wood is in commercial demand. A species for which there is no demand will not be extracted. If the commercial timber volume per unit area of forest is low then logging activities will probably be uneconomic. High transport costs are also a constraint. Demand for tropical hardwood comes from both home and overseas in about equal proportions, so logging should not be seen as being merely a matter of overseas exploitation of tropical countries.

Selective logging

A response to species abundance

The great species diversity of tropical rain forests presents difficulties to foresters and timber traders as the timber trade is species oriented and only a few of the many tree species are commercially marketable. Indonesia, for example, has about 3000 species of trees but only 107 are utilized. Two thirds of the commercial timber volume is represented by 20–40 species in Africa and 30–50 species in South America. Managing species-rich tropical rain forests is not easy and owing to the small number of commercial species they contain the kind of wholesale clearance common in temperate forests is totally unsuitable here. So selective logging, which involves removing a few good-sized trees of commercial species and leaving the rest behind, is the almost universal management system, although clearfelling has been practised on a few concessions in Papua New Guinea and Colombia (Webb and Higgins, 1978).

The volume of timber removed per unit area of forest in selective logging – the commercial cut per hectare – varies, with the commercial timber content, from country to country and site to site, from as high as 90 cubic metres in the Philippines to as low as 5 cubic metres in Brazil. This compares with a total timber volume per hectare averaging 258 cubic metres in Africa, 226 cubic metres in Asia-Pacific and 156 cubic metres in Latin America. Productive tropical moist forests, i.e. those in which logging is not prevented by poor tree quality, inhospitable terrain, or strict conservation protection, had an estimated 120,000 million cubic metres of timber in 1980 but only 8,000 million cubic metres of this was commercial (Lanly, 1981). Southeast Asia's dominance of the tropical hardwood trade is partly explained by the high cuts per hectare resulting from the high density of species of the Dipterocarp family with its valuable timbers. Different species in this family have similar wood properties and can be sold in groups as composite timbers: the woods of 200 commercial species in 11 genera in Southeast Asia are grouped as 10 major timbers (Grainger, 1986). Forests in Africa and Latin America are sometimes dominated in their composition by trees of the family Leguminosae (Gentry, 1986) but their leading timbers belong to less common families, such as the Meliaceae (which includes the genera *Khaya* in Africa, and *Swietenia, Carapa* and *Cedrela* in Latin America). For many years there has been criticism of the small number of species logged in tropical rain forests. While low intensity logging has a low impact on the forest a larger area has to be logged for a given volume of timber than if the forest were clearfelled. The obvious answer is to utilize more of the lesser-used (or secondary) species currently left behind in the forest because there is no market for them (Plumptre, 1974). But, as discussed in more detail in Chapter 7, despite the potential of many lesser-used species and strenuous scientific and marketing efforts the rate at which new species are accepted by the market is quite slow. Tropical timber traders are conservative, but that it is not the only reason. Many trees in Amazonian forests, for example, are often small in size, with high densities and silica contents that make sawing difficult.

Selective logging systems

Selective logging should not be a haphazard affair but a system of management relying on careful extraction followed by natural regeneration for future harvests. In most countries government forestry departments specify logging methods with a code of regulations that define the selective logging system most appropriate for local forests. Forestry departments are usually responsible for forest management because governments either own, or control the rights to exploit timber in most of the remaining tropical rain forest, though in some South American countries loggers can work on private estates, some of which are owned by timber companies. Sadly, granting rights to log so-called public forests can, as with minerals, often override the traditional rights of indigenous peoples unless the government is sufficiently enlightened to recognize these. In Papua New Guinea the traditional tribal system of land tenure is still legally valid and loggers there have to operate under permits from local people.

   In the majority of tropical rain forests logging is undertaken not by government foresters but by contractors working for individuals or companies awarded concessions by the government to exploit specific areas of forest. These 'concessionaires' are bound by their contracts to follow that country's selective logging system, the regulations of which specify the minimum diameter of trees that may be felled, the minimum number of trees to be left in the forest, and the time between successive harvests from the same patch of forest. Normally the second harvest is 30 to 40 years after the first, and since the trees to be logged then will already be in the forest at the first logging it is important that enough small diameter trees remain in the forest to provide the second harvest and that trees below a given diameter are not removed.

Selective logging systems vary slightly from country to country. In Indonesia, the minimum diameter is 50 cm, at least 25 trees of over 35 cm diameter at breast height (dbh) must be left in the forest, and the second harvest is 35 years after the first. In East Kalimantan forests this means that between 3 and 20 trees are felled per hectare (Kartawinata et al, 1984). In the Philippines the minimum diameter is 70 cm dbh, 70 per cent of trees between 15 and 65 cm dbh and 40 per cent of trees between 65 and 75 cm dbh must be retained, and the harvest cycle is 30–45 years. In Malaysia, the minimum diameter and harvest cycle are 50 cm dbh and 30 years respectively in Peninsular Malaysia, and 48 cm and 25 years in Sarawak. Instead of selective logging, Sabah uses a modified version of the Malaysian Uniform System (see Chapter 7) that requires heavy felling and regeneration on a long rotation of 60 to 80 years (Burgess, 1989). African harvest cycles are slightly shorter than those in Asia with minimum cutting diameters of 100 cm dbh for prime species in Liberia and Ghana, 80–100 cm dbh in Cameroon, 70 cm in Congo and Gabon, and 60 cm dbh in Ivory Coast (Rietbergen, 1989).

The advantages of selective logging are that a second harvest can be obtained sooner than if the forest were clearfelled and had to regenerate completely from bare ground, the regrowth of vines and fast growing non-commercial trees is reduced, and the impact on the forest ecosystem minimized. On the other hand, small or non-commercial trees can be damaged during logging (however carefully it is carried out), there is a risk that the population of commercial species will be depleted by continued extraction, and a larger area of forest has to be exploited per unit volume of timber than if clearfelling were employed. Logging damage is discussed more fully below, and the wider environmental impacts of logging in Chapter 6.

How concessions operate

Concession agreements give the holders rights to exploit timber in specific areas of forest, usually for 20–25 years, but with no guarantee of renewal. Agreements typically define the boundaries of the concession, the fees and taxes to be paid, standards for the construction and maintenance of logging roads, the annual rate of logging (by volume and/or area) over the concession period, the need for a full preliminary forest survey, and an obligation to protect forests after logging. Concession areas vary, averaging 98,000 ha in Indonesia,

40,000 ha in the Philippines and under 15,000 ha in Sarawak (Burgess, 1989). Concessionaires pay various initial and annual fees relating to the size of the concession and the overall volume of timber in it, annual royalties on the volume of timber harvested, taxes on timber that is exported, and a range of other fees (Gray, 1983). The last group might include a fee intended to pay for future restocking of logged forests, but the proceeds might not actually be used for that purpose.

Usually concessionaires know little of the commercial value of a concession before tendering for it and may well carry out the first timber inventory. Considerable investment is needed to build logging roads (and bridges) to access the forest and transport logs to the nearest market. Trees are felled with a chain saw (common in Southeast Asia) or hand saw (still used in Africa and Latin America). After the branches are removed the logs are hauled, usually by crawler tractors, along rough ('skid') trails to the nearest logging road and then taken by truck to the main logging camp. There they are sorted and graded before shipping by road or river to the nearest port or mill. The cost of road transport is generally six times that of river transport so good planning of logging activities, road routes and logging camps is vital to minimize costs.

Logging in practice can differ a lot from that required by regulations in the selective logging system and concession agreement. Many concessionaires remove more timber than they should, breaching the limits on minimum diameter and minimum number of remaining trees, and sometimes exploiting logged forest again before the concession agreement ends (Andel, 1978; Sastrapradja et al, 1978). Some forests in Ivory Coast have been logged three times in 10 to 15 years (Rietbergen, 1989) and not one of nine concessions studied in East Kalimantan had the required number of trees left behind for the second rotation (Soekotjo, 1977).

Careless logging also causes excessive damage to remaining trees, prejudicing the second harvest. Some damage is inevitable, for neighbouring trees are often linked together by vines so felling one tree can bring down or damage others. According to one rule of thumb, for every tree logged, a second tree is damaged and another is damaged but later recovers (Whitmore, 1990). Studies in the Philippines and Sabah found that up to 40 per cent of the residual stand was damaged, and in Indonesia the proportion was as high as 50 per cent (Weidelt and Banaag, 1982; Nicholson, 1979). Careless building of roads and trails leads to excessive soil erosion and deforestation: on well planned concessions only 10–14 per cent of the forest should be cleared for this reason but elsewhere it can rise to 50 per cent (Cacanindin et al, 1976; Nicholson, 1979).

Many concessionaires also fail to protect logged forests properly. Although they are supposed to ensure the forest can be logged again in 30 to 40 years time they have no assurance that any investment they make will lead to future income: their agreement ends after 20 to 25 years and there is no guarantee of renewal. They also know that local or migrant people may enter the concession and clear forest to grow crops, removing trees intended for the second harvest. This influences even the most conscientious of concessionaires: protecting logged forest from illegal clearance could be very expensive in large concessions.

Likewise, while it is in the interests of concessionaires to see that logging roads are properly surfaced and drained – otherwise log shipments by truck would be disrupted and revenue reduced – government road specifications are often circumvented to save money. Poor quality roads become a focus for soil erosion and gully formation after logging has ended, with severe environmental impacts.

In contrast to the mechanized techniques and large concessions common in Asia, logging in Brazilian Amazonia was until recently carried out by hand saw on a much smaller scale. With fewer commercial species and limited road access to dryland forests most of the logging was in seasonal swamp forests and performed by river dwellers rather than mechanized logging gangs (Carvajal, 1978). This changed during the 1980s as the new highway network and continuing deforestation for agriculture improved access to forests in Rondônia, Acré, Mato Grosso and Roraima states and allowed more intensive logging, the wood being sawn locally and then exported (Browder, 1989). In Amazonia agricultural colonization generally preceded logging, whereas in other regions the reverse was the case. The most popular Amazonian timber is virola, a lightweight structural timber (Table 3.2). Mahogany and andiroba (which has a similar wood) are also sought after but are only found scattered in the forest.

Is selective logging sustainable?

In the late 1980s the sustainability of tropical rain forest logging became a major issue for environmentalists in developed countries, who called for all tropical hardwood to be produced in sustainably managed forests within the space of a few years (1995 was the deadline cited recently by the Worldwide Fund for Nature). But this concern was not new, as from the late 1970s the governments of producing and importing countries had recognized the need to work together to improve forest management, and this led to the International Tropical Timber Agreement (see Chapter 9), which came into force in 1985. But how sustainable is logging today and how fast can the degree of sustainability be increased?

What is sustainable management?

An area of forest is said to be managed when timber harvesting is controlled to satisfy given objectives. Often the very use of the word 'management' implies the intention of sustainability but in the past tropical foresters often used it rather lazily just to indicate that forests were being logged. So the term 'sustainable management' is now employed to describe a specific form of management whose explicit objective is to maintain timber yields at satisfactory levels, taking into account the ability of the forest to regenerate after logging, assisted as necessary by human intervention. Yields are judged satisfactory on the basis of economic and biological criteria. The yield at the second harvest will usually be lower than the first, but thereafter yields should stay fairly constant if the logging system has a sound biological basis.

Forest management should not be equated merely with silviculture, which is a set of techniques to give a preferred mix of commercial timber species at the next harvest by manipulating natural regeneration after logging or assisting it by planting trees (see Chapter 7). Duncan Poore (1989a) showed the wide scope of management by listing five different levels:

1. pure protection in anticipation of future logging;
2. logging forest and then leaving it in a protected state to regenerate naturally before the second harvest, i.e. selective logging;
3. logging, followed by natural regeneration with limited human intervention;

and two silvicultural options:

4. logging, followed by intensive intervention to manipulate postharvest natural regeneration;
5. logging, followed by manipulation of natural regeneration and planting of desirable species needed for the second harvest.

Forest protection is required in all cases.

Whether or not management is sustainable may be assessed by various criteria. These include the presence of a written 'working plan', a basic tool of professional foresters that specifies in advance the forest operations to be undertaken and the expected yields. The biological and economic bases for sustainability are determined in general terms by the selective logging system, local forest conditions and the concessionaire's business plan, which takes into account both operational costs and government taxes and incentives. The extent of official monitoring, regulation and protection is also important, since actual logging activities often depart from government regulations and the concession agreement. Forestry department personnel should monitor concessions to check on this and impose sanctions if regulations are not complied with. However good forestry operations are, sustainability is threatened if there is illegal forest clearance by outsiders, so effective protection measures are vital, though permanent national forest reserves are not essential in the early stages of forest exploitation.

The International Tropical Timber Organization has been preparing a more formal set of criteria to measure the degree of sustainability in practice. At national level these include: the legal status of the country's forest resources, the continuity of flow of forest products, the level of environmental control of forestry, the socio-economic effects of forest management and the institutional framework that supports it. Local criteria include: the degree of forest resource security, the continuity of timber production, the conservation of flora and fauna, the level of environmental impact, the pattern of economic benefits, and provisions for planning and adjusting practices on the basis of experience. This list is currently being refined and supplemented by practical indicators for each criterion (ITTO, 1990c, 1992; Cassells, 1992a).

The sustainability of logging systems

The overall sustainability of selective logging depends on the sustainability of logging systems and how they are implemented in practice. Our knowledge of regeneration processes in tropical rain forests is so uncertain that there must be some doubt about present systems. Most are less than a century old and according to one expert a firm judgement on sustainability is impossible until a patch of forest has been logged twice and in at least its third rotation (Poore, 1989a). There is no guarantee that predicted yields of desirable species at the second and succeeding harvests will be obtained given the harvest cycles of 30 to 40 years now common (Ashton, 1978). Ultimately, second harvest yields depend on whether enough trees of sufficient size remain in good condition after the first logging. Some experts claim that they will (Setyono, 1978) but others disagree

(Weidelt and Banaag, 1982). The only way to be sure is to monitor yields under experimental and actual concession conditions but this will take time (Palmer, 1989). The list of commercial species is always changing, so while in 40 years time a forest may not give predicted yields of presently commercial species it could give good yields of species then in demand. All we can say at the moment is that present selective logging systems may well be the best available but there is scope for improvement. However, because of the need for long-term monitoring and research, current faults in logging systems could take some time to rectify.

The sustainability of logging practices

Even if a selective logging system is sustainable under controlled experimental conditions it will not necessarily be so in practice, owing to the overlogging, excessive forest damage and poor protection described above. Loggers can get away with this because forestry departments usually station too few forest officers nearby to monitor and regulate their activities. Those that are there are poorly paid and vulnerable to being bribed not to notice when more trees leave the forest than is allowed. Monitoring is made harder because most government forestry departments lack independent data on the initial timber content of concessions and have to rely on surveys by concessionaires themselves. Lack of proper forest protection is a serious threat to sustainability but for a concessionaire to safeguard for twenty years often tens of thousands of hectares of logged forest from illegal clearance by encroaching cultivators and opportunistic logging by timber poachers is a formidable and expensive challenge, especially if there is no guarantee that the cost will be reimbursed by the proceeds from a second harvest.

Ideally, to help ensure sustainable management all productive forests should be inside properly demarcated, legally binding and fully protected reserves. But only a small fraction of all tropical rain forest currently satisfies this criterion and showing reserve boundaries on forest maps at headquarters does not mean they are well protected on the ground. This is true for reserves in Ghana, Ivory Coast, Cameroon and Liberia (Rietbergen, 1989)[6], and some forest reserves in Peninsular Malaysia, Sabah and Sarawak have even been officially converted to agriculture. Indonesia has classified forests into long-term land use categories[7] but does not yet have forest reserves as such. Nor do the Philippines, Thailand and Papua New Guinea – where the traditional tribal land tenure system would make reserves impossible anyway (Burgess, 1989). Bolivia, Peru and

Ecuador have significant reserves but they are poorly protected. That is also probably the case for the limited reserves in Brazilian Amazonia. In Trinidad and Tobago, the only Latin American country where forest management is regarded as sustainable, only 45 per cent of all forest is in reserves (Synnott, 1989).

It is unrealistic to expect reserves to be established in the early stages of forest exploitation, as this requires the consent of other government departments, the presence of sufficient forestry personnel on the ground and, most importantly, the active cooperation of local people. All these ingredients take time to assemble. Foresters have learned from bitter experience that keeping people out of forests where they have traditionally hunted and gathered, and imposing strict policing with fines and jail sentences for those who encroach into the forest, is a recipe for disaster. Antipathy between foresters and local people leads to more deforestation, not less. So working with local people is the only way to achieve long-term forest protection, but it can take years to build trust and agree practices acceptable to all parties.

The present extent of sustainable management

According to data from FAO, which requires managed forest to be in permanent reserves with management operations listed in detailed working plans, less than 1 per cent of all productive tropical rain forest (8.4 million ha) was under any form of management in 1980 (Table 3.8) (Lanly, 1981, Grainger, 1984)[8]. Most managed forest in Africa was said to be in Ghana and Uganda, but political and economic difficulties at the time meant that the true extent of management in these two countries was questionable. Management in Asian forests was concentrated in Burma and Malaysia (mainly Sarawak). The only managed forests in Latin America were in Trinidad and Tobago and few forests in that region have ever been managed.

Table 3.8 *Area of managed forest in the humid tropics in 1980 (thousand ha)*

Africa	1,658
Asia-Pacific	6,753
Latin America	14
Total	8,425

NB Excludes tropical moist deciduous forest in India
Source: Grainger (1984), based on Lanly (1981)

Controlling tropical deforestation

The FAO data did not give a complete picture of the extent of sustainable management, however. Not all the forest within reserves would have been managed sustainably, but a lot of forest outside reserves was probably under partially sustainable management. A more recent assessment, commissioned by the International Tropical Timber Organization, concluded that sustainable management was essentially confined to Trinidad and Tobago and (part of) Ghana, former management systems elsewhere in Africa had been abandoned, and Asian forests under concession agreements were, except for Papua New Guinea, nominally under management, though there was a great difference between theory and practice (Poore, 1989b).

The sustainability of tropical rain forest management is therefore limited by shortfalls in both selective logging systems and their implementation. It is easy to list sustainability criteria but harder to label forests as being either sustainably managed or not. Duncan Poore concluded that: 'it is not yet possible to demonstrate conclusively that any natural tropical forest anywhere has been successfully managed for the sustainable production of timber' (Poore, 1989a). Poore's view was that the main constraints on sustainability were organizational, not technical, relating to the implementation of logging systems rather than the systems themselves. So there is scope to improve sustainability without waiting for the results of research that could lead to improved systems. Better monitoring, regulation and protection of concessions could lead to great benefits but in practice these are unlikely to happen overnight.

Logging impacts and sustainable management

Two issues have dominated this chapter: the impact of logging on tropical rain forest and the sustainability of forest management. Because of the great species diversity in tropical rain forest logging is mainly selective and degrades the forest rather than causing deforestation as defined here. Since its impact on the forest differs from that of agriculture it should be treated as a separate phenomenon, involving, besides a temporary disturbance of the forest canopy, consequential damage to the remaining stand and some temporary clearance for roads, etc., both of which are exacerbated by poor logging practices. Selective logging has significant ecological and environmental impacts on the forest, discussed further in Chapter 6, and can also be an indirect cause of deforestation if farmers use logging roads to gain access to forest.

Harvesting tropical hardwood is a legitimate way to exploit tropical rain forests as it allows forest cover to be maintained. But it is important that forest management becomes more sustainable, both from the point of view of treating forest as a renewable source of timber and maintaining forest cover for its many benefits. At present, very little tropical rain forest is under fully sustainable management, but there is great potential to improve on this, and some possible strategies are explored in Chapter 7. Halting logging would not have much of an immediate impact on deforestation rates but making forest management more sustainable will reduce the amount of forest degradation caused by logging. Based on past experience, however, achieving sustainability will take time, because it is not just a matter of introducing better techniques. The organization of national forest management also needs to be improved and this is something which, given the large areas involved, can only happen gradually.

4

People, policies and forests

The previous two chapters described how deforestation and logging occur in tropical rain forests. This chapter goes into the matter a little more deeply, looking first of all at *why* they take place as they do. Land use change and logging do not happen by themselves. Demand for food, other farm products and minerals means that land is needed to produce them and this often requires deforestation; forest is logged to supply demand for wood. These demands in turn can be traced to more fundamental causes, which include social and economic factors and government policies (Table 4.1).

Population growth and economic development are the main driving forces behind deforestation and contribute also to rising demand for tropical hardwood at home and overseas. Government policies affect land use change directly, as well as indirectly through their effect on population and economic activity. The spread of deforestation is influenced by environmental factors, ease of access to forest, and the sustainability of land use, while the international spread of logging activity is governed by economic considerations such as the costs of logging and transport and also by government policies.

Another crucial question tackled here is what controls long-term trends in forest cover and logging in particular countries. We can point to various possible mechanisms that could control deforestation: investment in more productive farming practices should mean that eventually less land is needed for farming so the pressures to clear more forest are greatly reduced; governments should gradually increase the protection accorded to forests; and poorly accessible forests should probably be free from deforestation. As to when these controls will come into operation there is no definite answer except to say that deforestation trends in each country are part of a general evolution in national land use patterns. Historical experience in the developed countries offers two alternative

Table 4.1 *Major underlying causes of deforestation*

---

1. Socio-economic factors
   a. Population growth
   b. Economic development
   c. Poverty

2. Physical and environmental factors
   a. Distribution of forests
   b. Proximity of rivers
   c. Proximity of roads
   d. Distance from urban centres
   e. Topography
   f. Soil fertility

3. Government policies
   a. Agriculture policies
   b. Forestry policies
   c. Other policies

---

scenarios: centuries of deforestation and logging reduced forest cover in the UK to just 5 per cent by the end of the nineteenth century, but in the USA deforestation was brought under control at a much earlier stage when only half of the original forest cover had been lost. These wider trends in land use will also affect the future of logging as they will determine how much forest remains to be exploited for timber, but the principal influences will be trends in wood demand, logging and transport costs, government policies and forest management. The quality of forest policies and management systems, and the ability to implement them over large areas, also evolve over time. This was the experience of temperate countries and there is no reason why tropical countries should be any different, though there is a question mark over the speed of evolution.

This chapter looks at all of these issues, building on the material in Chapters 2 and 3, and is in two main parts. The first part begins by reviewing the underlying causes of deforestation and then looks at historical trends in forest cover, mechanisms that could bring deforestation under control, and the impact of deforestation on forest dwellers. The second part discusses the economic factors and policies that determine the spread of logging, how logging is carried out, the development of forest industries and the role of tropical hardwood in world trade.

The underlying causes of deforestation

In Chapter 2 the changes in land use leading to deforestation were explained in various ways. Here we highlight six main underlying causes: population growth, economic development, ease of access, environmental factors, government policies and land use sustainability.

Population growth

The rapid rise in population in tropical countries is one of the major forces driving deforestation, as more people need more food and hence more land on which to grow it. When most unused land is covered by forests, as is generally the case in the humid tropics, deforestation is inevitable if agriculture is to expand. Rising demand for food could be met by greater agricultural productivity, of course, but the poor economic circumstances of most tropical countries make this difficult. Even if productivity does increase, there will probably still be a need for some deforestation in the early phases of development.

Populations are rising rapidly because economic development brings about a general improvement in health care and quality of life, which lowers national mortality rates, but fertility rates do not fall until much later when people feel secure enough to reduce their need for dependants. This theory of population growth, called the 'demographic transition', assumes that population growth rates will decline eventually, but is still controversial in some quarters (Notestein, 1945; Caldwell, 1976; Teitelbaum, 1975; Ness and Ando, 1984).

Population growth leads to deforestation in two main ways. A rise in overall national population, i.e. in both rural and urban areas, increases demand for food, some of which is satisfied by the market system and some by subsistence farming. This link is actually supported by a statistical correlation between the rates of deforestation and population growth in the 1970s (Grainger, 1986; see also Allen and Barnes, 1986)[1]. Deforestation is also caused by local population increases, which may result from either inherent growth or immigration. In rural areas still highly dependent on subsistence food production much of the extra food needed to support higher populations has to be supplied by either greater cropping intensity (i.e. shorter shifting cultivation rotations) or an expansion in farmland and hence deforestation. Urban population growth should depend much more on the market system and favour the expansion of permanent agriculture. It also leads to deforestation in the immediate vicinity for fuelwood, building materials and land for settlement. A variety of migration patterns may be observed, involving movement between different rural areas, from rural areas to urban areas, and from urban to rural areas as people who came to towns in search of jobs but were unsuccessful return to the countryside, often as encroaching cultivators. In addition, some people keep their rural base but work in urban areas, either part- or full-time to support their families, while others are based in urban areas but travel out into the countryside at weekends, or occasionally for longer periods, to tend farm plots.

Economic development

Population growth is part of the wider process of social and economic development seen in the tropics over the last 40 years. Economic development, as formally defined, should increase a country's material prosperity and social welfare, reduce poverty and inequalities between different sections of the population, improve the quality of life of individual citizens by provision of better health, education and other services, and increase the quality and quantity of economic output by better access to technology. It depends on economic growth – the rise in the average rate of provision of material goods and services, measured by the Gross National Product or GNP – but is not identical to it, for if the extra income is poorly distributed among a country's population the lot of the average citizen is little improved (Simpson, 1987).

Economic development, however slowly it occurs, involves major changes throughout society, many of which impinge on deforestation. As average income rises each person might be expected to increase their food consumption, so demand for food would increase even if population did not. Many people desire a more 'civilized' lifestyle, which may entail reducing shifting cultivation rotations or swapping shifting cultivation altogether for permanent, sedentary, cultivation. National economies become more closely integrated into the world economy and governments promote the exploitation of timber and other natural resources and the expansion of cash crop plantations to raise foreign currency to pay for national economic development. They also often wish to establish control over far flung territories: as stated earlier the recent settlement of Brazilian Amazonia was in part an attempt by the government to prevent an anticipated invasion from neighbouring countries. All these factors favour more deforestation.

An important consequence of economic development is the encroachment of the market economy into traditional rural subsistence societies still largely dependent on barter rather than cash as the medium of exchange. This can lead to the disintegration of traditional cultures that previously exerted the controls needed to ensure sustainable use of common property resources such as tropical rain forests. When cultures are degraded so too are these social controls on land use, the forests become more like open access resources (Pearce and Turner, 1990) and deforestation is allowed to spread. Local societies then lack the necessary cohesion to manage things as they did in the past, but governments rarely have sufficient employees in wilderness areas to manage forests properly themselves, often failing also to recognize legally the traditional land rights of indigenous peoples.

Other aspects of economic development can actually slow down the rate of deforestation. Rising income should allow greater investment in fertilizers, high yielding varieties, tractors and other inputs that will make agriculture more productive. If average food yields increase, less land is needed to grow the same quantity of food and this should reduce deforestation rates. The national economy, formerly dominated by agriculture, should diversify to include significant activity in the industrial and service sectors, and as rural people migrate to the cities and the proportion of urban dwellers rises the overall pattern of land use changes, farming becomes more intensive and the proportion of people employed in agriculture declines. In conclusion, we can see that economic development therefore has contradictory effects on deforestation, on the one hand promoting it and on the other controlling it (Grainger, 1986).

Poverty, landlessness and debt

But investing in the future is not easy. Many countries have difficulty in ensuring that economic growth outpaces population growth. Often they are unsuccessful so the gains in average income are minimal and less money is available for long-term investment. The wealth that is generated is usually not spread evenly among the population and most small farmers lack the means to invest in more productive and sustainable forms of agriculture that would reduce deforestation. The so-called 'Green Revolution' in the 1960s made high-yielding crop varieties, capable of yield rises of up to 300 per cent, available to all farmers in principle, but in practice often only the larger, richer farmers benefited. Many smaller farmers who tried to join in the revolution got heavily into debt, owing to the high costs of seeds, fertilizers and pesticides, often losing their lands if they could not repay their debts.

If poverty in a relative sense is an underlying cause of deforestation in that it prevents investment in more productive farming practices, then more extreme poverty leads to deforestation in a far more basic way, as landless people migrate from cities or other rural areas to clear forest simply to grow enough food to survive. Economic inequality and unequal access to land is deeply entrenched in many countries and, as discussed in Chapter 2, it may be traced back hundreds of years to colonial times. Too often economic growth widens, rather than narrows, the gulf between rich and poor in developing countries. In the humid tropics the forests are the last resort of the landless. Poverty and landlessness, and the deforestation associated with them, are likely to remain problems for some time to come unless governments modify their development strategies.

In recent decades the governments of developing countries borrowed a lot of money from overseas banks and international agencies to invest in future development. But as commodity prices fell during the 1980s many countries could not repay the interest on these loans and lapsed increasingly into debt. By 1986 the total external debt of developing countries had reached $1000 billion, of which three quarters was owed by governments. Some commentators blamed foreign debt for exacerbating deforestation, claiming that governments consciously increased the rates of logging, mining and plantation establishment to increase foreign earnings to repay their loans, while at the same time reducing government spending, so that poverty-stricken people had to clear forests to survive. This explanation is convenient, became the accepted norm in some quarters in the late 1980s and was given superficial credence by the use of 'debt-for-nature swaps'. However, deforestation was occurring well before the debt crisis and there is no hard evidence that debt had a significant new effect on deforestation rates. Indeed, without investment in more sustainable land management the deforestation problem will certainly be a lot worse in the future.

Ease of access

The spread of people and farming techniques into forested areas is influenced by the distribution of forest in relation to populated areas and the ease of access by roads and rivers. Roads are still few and far between over large areas of the humid tropics. So rivers often remain the chief means of communication and even today permanent cultivation tends to be concentrated close to waterways. Building new roads into an area can therefore have a huge impact on land use by allowing large numbers of people to migrate there. This happened to startling effect with the building of highways in Brazilian Amazonia. Another example is in the Indonesian province of East Kalimantan in eastern Borneo where Balikpapan and Samarinda, respectively the commercial centre and capital of the province, were not linked by road until the 1970s (Figure 2.2). That highway remains one of only a handful in an area almost the size of Great Britain. After it was completed migration began and devastated wastelands left behind by encroaching cultivators stretch on either side as far as the eye can see. If highways link up with logging roads even deeper incursions into the forest are possible, leading to 'logging-deforestation feedback' in which the spread of logging catalyses that of deforestation. Roads also connect farmers to urban markets and promote cash crop cultivation if transport costs are satisfactory. Settlers along the TransAmazonian Highway were upset when the Brazilian government failed to build promised feeder roads to their farm plots, so they could not get their crops to market easily. Settlement along BR 364 in Rondônia was partly explained by the good links it gave to major urban markets in southeast Brazil.

Environmental influences

Environmental factors such as topography and soil fertility also influence the spread of deforestation but in less predictable ways. Lowlands should ideally be more attractive for colonization than hilly areas – if they are not swampy – but though in Southeast Asia farming generally began in the humid lowlands and later moved to the highlands, in Latin America the opposite trend was seen. A study of deforestation patterns in part of Costa Rica found the highest proportion of forest clearance on steep lands with gradients of 31–45 per cent (Sader and Joyce, 1988), but there are instances in Southeast Asia where shifting cultivators from hilly areas have migrated to the lowlands to clear forest and lead more sedentary lives.

Soil fertility also has ambiguous effects. Fertile areas are likely to be chosen for tree crop plantations and other capital intensive agricultural projects, but settlements in the Indonesian Transmigration Programme were often poorly sited in the past, and both official and spontaneous settlement along the TransAmazonian Highway was less than hoped for because it was routed mainly through areas with infertile soils. Meanwhile, spontaneous settlers flocked up BR 364 to Rondônia where soil fertility was better. Elsewhere, poverty and landlessness may force people to clear forest regardless of fertility, even if they can only crop the land for a few years before moving on.

Unsustainable land uses

Land uses do not just have a passive role in deforestation. As shown in Chapter 2, the size of clearances varies with the land use replacing forest, from a few hectares a year cleared by a shifting cultivator to many thousands of hectares over a short period for a cattle ranch. But central to the future of deforestation is the sustainability of land use in deforested areas. Owing to the generally poor fertility of soils in the humid tropics unsustainable farming is a major hazard and a cause of deforestation in its own right, for if yields decline, sometimes to the point that land has to be abandoned, more deforestation is necessary to maintain overall food production. Land uses can become unsustainable if introduced on to sites to which they are unsuited, as seen in poorly planned resettlement schemes where forest on poor soils is cleared for permanent field crop cultivation and acceptable yields only last for a few years. The sustainability of one land use can also be affected by the impact of another, as when permanent cultivation expands into an area formerly used for shifting cultivation, the fallow period is cut to accommodate the smaller area available, and the sustainability of shifting cultivation is threatened. The challenge to agricultural planners is to ensure first that the most productive agriculture is concentrated on the best sites, leaving poorer sites to less intensive uses, and second to minimize the interactions between different land uses that could prejudice their sustainability. Deforestation will remain a major problem unless greater care is taken in this respect.

The role of government policies

So far deforestation may have seemed a rather mechanistic response to socio-economic change but there is a considerable element of choice involved and this is particularly evident in the role of government policies. The easiest influences to identify are where governments specifically promote the expansion of certain land uses and the building of highways as part of their national development strategies. Amazonian deforestation rates over the last 20 years, for example, would have been much lower had the Brazilian government not heavily subsidised cattle ranching and other large development schemes, built a new highway network and linked it with planned resettlement schemes. Many other governments are undertaking or promoting large-scale agricultural development projects, and there is nothing wrong in this necessarily, for planned deforestation can lead to deforested land being used in a sustainable way that minimizes environmental impacts. But problems arise when governments do not take their planning role seriously, and introduce land uses on to sites to which they are unsuited or undertake highway construction and other projects with the potential for serious environmental impacts. Too often development is seen as an end in itself and the role of natural resources in development strategies is simply as a source of short-term income.

Governments also influence the rates of growth in population and economic activity, the major underlying causes of deforestation, but the link with deforestation is less evident because it is indirect and the policies are more general in scope than those relating to agricultural or highway development. They also vary a lot from country to country. Some governments, such as that of Malaysia, promote population growth in the national interest; others try to slow it down by supporting family planning; while another group of governments, including those of Indonesia and Brazil, resettle part of the existing population to less crowded areas. Whatever the population policy may be there is a general tendency for governments to ignore the impacts that their social and economic policies have on land use or the environment.

Why is the role of government often so detrimental? Partly because their policies have serious limitations, partly because of poor implementation. In most cases the choice of policies reflects a technocratic approach to development on the part of government and a wish to see their countries develop, whatever the cost. But the overall shape of policy, and the beneficiaries of economic growth, are also determined by the distribution of power in the country. It is a mistake to think that all people in developing countries are poor: severe inequalities in wealth and power are common, even within villages, and there are disparities too between rural and urban dwellers. Though governments have a duty to ensure that the benefits of economic growth are shared as fairly as possible among all citizens this is rarely achieved. The interests of elite groups are likely to be dominant in policy formation: large corporations wield power just as in industrialized countries, and leading companies, including those involved in natural resource exploitation and other development projects, are often controlled by relatives of leading politicians or members of the military. The armed forces have a major say in policy formation in most countries and are a prime beneficiary of government spending and decisions, even being awarded logging concessions in some cases.

The influence of major landowners is such that policies tend to favour large farming estates rather than small farms, and in countries like Brazil and the Philippines with grossly inequitable distributions of land ownership governments have failed time and time again to reform land tenure to redistribute land from rich to poor. This problem is particularly acute in Latin America and the Philippines where land tenure systems have been biased in favour of large estates since the Spanish and Portuguese colonial period (see below). But while landlessness is a major social problem in the humid tropics and an important cause of deforestation, redistributing land could lead to even more deforestation since the recipients would naturally want to clear forest on their new properties. Fear of redistribution actually caused landowners in Brazilian Amazonia during the 1980s to cut down forest on their estates to give the impression that the land was being put to 'productive use'.

Lack of attention to promoting broadly-based agricultural development is symptomatic of a far wider problem. In the present developed countries urban-industrial expansion occurred only after a strong agricultural base had been established, but many tropical countries are trying to circumvent this stage, promoting urban-industrial development to try to accelerate economic growth. This also reflects political realities, for the survival of governments often depends on keeping urban populations happy. Expanding urban services rather than satisfying the needs of rural people may allow governments to stay in power but it perpetuates rural poverty. Keeping food prices low appeases urban people and maintains political stability but gives farmers no incentive to invest in more productive practices. As countries become more democratic this should gradually reduce the concentration of power and the scope for corruption, but the urban bias will take longer to disappear.

However bad for the environment many government policies are, even good policies fail due to inadequate implementation. Lack of personnel is a major constraint. Loggers can flout the rules of selective logging systems if there are too few government foresters in the area to monitor their activities and those that are there are poorly paid and susceptible to bribes. A shortage of trained personnel causes problems at higher levels, as top scientists may have to undertake full-time management duties rather than urgently needed research, e.g. in forestry or land evaluation. The organizational capability of government ministries is often inadequate for the task of administering large areas of lands and forests. Technical limitations are evident in a lack of equipment, and what is available may not function properly because spare parts or repair staff are in short supply. Funding shortages also limit the availability of petrol for official vehicles, so that foresters, for example, could be unable to monitor and regulate logging and curb deforestation as well as they would like.

Historical trends in tropical land use

Current trends in tropical deforestation should be seen as a continuation of a long process of land use development, though past changes in the humid tropical zones have been generally less than those in drier or more seasonal areas, due mainly to environmental factors. Since land use is closely linked to social conditions the patterns of deforestation seen today also reflect how each country's society and land use have evolved, interdependently, over time.

Hunter-gathering origins

Human beings only became farmers 10,000 years ago. For most of our history we were hunter-gatherers. Though rare nowadays, the hunting of animals and the gathering of fruits and other plant parts is still practised by tribal forest-dwelling peoples, such as the Pygmies of the Congo Basin and some of the Penan of Borneo. Other peoples, like the Yanomami of Amazonia, are shifting cultivators as well. An individual Penan, for example, gathers an average of one wild sago palm each week to use for all purposes, from food to thatching material (Hanbury Tenison, 1979). Hunter-gatherers have a minimal impact on the forest, leading mainly nomadic lives with low population densities of typically 0.25 persons per square kilometre.

The rise of agriculture and civilization

It would be wrong to think of hunter-gathering as the dominant traditional land use in the humid tropics, for the earliest known instance of shifting cultivation there involved the growing of two root crops, taro and yam, in Papua New Guinea as early as 8000–7000 BC. It is significant that this was about the same time that wild cereal grass cultivation started in dry upland areas in West Asia, the site of the world's earliest known settled farming villages. Rice growing by shifting cultivators in Southeast Asia started between 5000 and 3500 BC, and the more sophisticated wet rice cultivation was introduced from China to Indonesia and the Philippines in the second millennium BC, spreading all over eastern Asia after 500 BC (Simmons, 1989). Agriculture probably did not start in the African humid tropics until the first millennium BC when the Bantu people began to grow yams and taro imported from Malaysia. However, it is thought to have originated independently in Latin America, with maize being grown in upland Central America by 5000 BC, and about a thousand years later in Amazonia (Bush et al, 1989; Roberts, 1989).

The achievement of food surpluses supported the growth of civilizations in some areas, e.g. in India (by 2500 BC) and the Andes (from 1000 BC onwards). The Sahara Desert blocked the spread of civilization from Egypt to the African humid tropics and the Mayans were the only civilization to flourish in the Latin American humid tropics. In Asia, on the other hand, Indian civilization was transferred to Burma, Thailand, Cambodia and Indonesia and, as David Wyatt has stated, 'it is not unreasonable to suppose that by two thousand years ago the peoples of Southeast Asia shared a common, distinctive and advanced civilization' (Wyatt, 1982).

The heritage of colonial rule

This was certainly the experience of the first Europeans to visit the region and led to a different pattern of colonization there than in Latin America, based initially on trade rather than occupation and subjugation. In most tropical countries, however, the natural evolution of land use was diverted in some way or other when they were colonies of Britain, France, Spain, Portugal, Belgium or Germany and the repercussions are still felt today.

The conquest of Latin America came first, starting in the Caribbean in the early sixteenth century. Shortly afterwards Spain occupied the Philippines but other European powers focused on building trading links with Asia, leaving formal colonial occupation until the nineteenth century. Trade between Europe and Africa was sparse before 1800 and its colonization took place last of all, in the latter part of the nineteenth century. By then Spain and Portugal's American colonies had already gained their freedom, but most colonies in Africa and Asia did not follow until after the Second World War. So the post-war period of economic development has also coincided with a new era of self-determination for many tropical countries (Fieldhouse, 1982).

There is much debate about the original intentions of the colonizing powers. One view is that their aim was to secure valuable raw materials for use at home and at the same time create new markets for their own manufactured products (Simpson, 1987). Another dismisses such general long-term strategies, favouring a more incremental, reactive approach (Fieldhouse, 1982). Whatever the true explanation, in bringing their civilizations to the tropics Europeans denied tropical peoples the right to develop as they would wish, a freedom only recently obtained by most of the former colonies.

This is most evident in the effect of colonial rule on the evolution of tropical economies and land use patterns. Natural resources became important trading commodities: the Spanish stole gold from Inca and Mayan temples in Peru and Mexico and then ran the gold mines with local slave labour; the Portuguese became wealthy from the spices they shipped from Southeast Asia; the Dutch mined tin on Java; and the British extracted teak from forests in Burma. In establishing plantations to produce sugar, coffee, cocoa and other cash crops, the colonial powers stamped their own imprint on land use and indigenous evolution of land use patterns was redirected. Considerable deforestation was caused, one notable example being that needed to turn Burma's Irrawaddy Delta region into a huge rice growing area. But although sugar plantations were established early on in the first Caribbean colonies, such as Cuba and Hispaniola[2], other important plant commodities, such as oil palm and rubber, were for many years collected from wild stands in tropical rain forests and plantations came much later.

The impact of plantations went far beyond land use patterns. The relationship between people and the land was changed too. Local people worked as labourers rather than managers and operations were controlled by companies from the colonizing country. A dual agricultural economy emerged, composed of subsistence agriculture on the one hand and a plantation economy on the other. In South America and the Philippines the present imbalance in land tenure may be traced to the large agricultural estates established long ago by the Spanish and Portuguese. For these and other reasons Piers Blaikie and Harold Brookfield (1987) have described colonialism as 'a revolutionary process ... that ... sought to terminate or divert former evolutionary trends.' Ironically, when tropical countries gained their independence they had little choice but to build on their colonial heritage to earn foreign currency to fund their economic development, hence the continuing expansion of the plantation sector.

Future trends in deforestation

How are tropical deforestation rates likely to change in the future? Is every country fated to lose all of its forest, or could deforestation be brought under control? If so, instead of the simplistic downward linear trend in forest area beloved of some commentators perhaps the rate of forest decline could tail off (Figure 4.1). Is this feasible, and if so how would it happen?

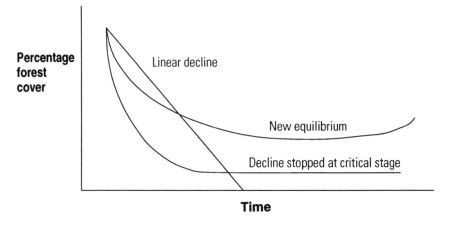

Figure 4.1 *Alternative scenarios for forest cover decline*

Most forested countries undergo at some point in their histories what this author has called a 'national land use transition' as their land use pattern changes from one dominated by forest to one dominated by agriculture (Grainger, 1986). Many countries in the humid tropics are now in the throes of such a transition. Most temperate countries, such as the UK, passed through it centuries ago. Farmers are the main protagonists in this and however hard foresters and conservationists try to protect forests they are usually not very successful. The transition may take hundreds of years to complete and leaves the country considerably less forested. Land use will continue to change afterwards, and forest may even regenerate or be replanted, but usually the percentage of national forest cover will never return to its original level.

During this transitional period the rise in demand for food will be met by a combination of increased farming intensity and expanded agricultural area. Each of these two options can be met in turn by either shifting or permanent agricultures, so the way in which farming practices evolve varies from one country to another. The relative rates of expansion of the two farming types are difficult to predict but shifting agriculture will tend to be preferred in poorer countries and permanent agriculture in richer ones. A general trend from shifting to permanent agriculture would, however, be expected everywhere. The ratio between areas under shifting and permanent cultivation in Peninsular Malaysia, for example, is much lower than in less developed East Kalimantan (Grainger, 1986)[3]. How far and how fast the trend will proceed is uncertain. Some experts believe in an automatic switch from shifting to permanent agriculture with socio-economic development (Boserup, 1965; Seavoy, 1973; Watters, 1960) but others are more sceptical (Blaikie and Brookfield, 1987). Even if shifting cultivation does remain widespread, permanent agriculture will probably take over the more fertile areas, become increasingly productive, and account for a growing proportion of all food and non-food crop production. Ideally, the need for new farmland will decline and so will the annual deforestation rate. National land use should approach a new equilibrium in which forest and farmland areas are again in balance, but with a far lower national forest cover than before. The national land use transition will have been completed.

Historical experience in developed countries offers two alternative scenarios for the transition. At the onset of European settlement 45 per cent of the USA was forest covered. Total forest area was the same as in Brazilian Amazonia today and by the 1850s the annual deforestation rate by agricultural clearance and logging was close to that in Brazilian Amazonia in the 1970s (Williams, 1989). Yet by the end of the nineteenth century large areas of low quality agricultural land in the east had been abandoned in favour of more productive land in the central USA, logging had been controlled and areas of forest protected for long-term timber production and conservation. The deforestation rate fell and the decline in national forest cover levelled off at about 25 per cent, though 70 per cent of remaining forest was in a degraded state. Before long national forest cover actually increased owing to regrowth on abandoned farmland. So in the USA a combination of rising agricultural productivity – with farming concentrated on the best lands – and protection of forests for timber production and conservation brought deforestation under control with a substantial proportion of the country still forest covered.

Contrast this with what happened in the UK, which had a much longer history of deforestation and logging, starting thousands of years ago. Government protection of forests began in Saxon times (AD 450–1066) but was never really effective, and the monarchy often connived in deforestation to enrich its treasury. Warning bells that forest resources had reached critical levels sounded in the seventeenth century but national forest cover continued to fall, reaching just 5 per cent by the end of the nineteenth century, and no serious attempt was made to halt deforestation and promote reforestation until after the First World War. Since then forest cover has risen steadily to 10 per cent due to a determined forest plantation programme. In the UK, therefore, forest cover fell to a very low level before government intervened to stop the decline (James,1981; Grainger 1981).

Which pattern will tropical countries follow? Available data support the notion that the decline in forest cover should level out, but with the amount of forest remaining likely to vary from country to country. A plot of national forest cover against population density (used here as an indicator of national development) for 43 countries in the humid tropics shows forest cover moving to a limit of about 18 per cent, but separate plots for the three tropical regions give limits of 15–20 per cent for Asia-Pacific and under 10 per cent for Africa and Latin America (Figure 4.2) (Grainger, 1986). In the Philippines, which has one of the highest population densities in the humid tropics, national forest cover now appears to be levelling off at about 24 per cent (Figure 4.3) (Kummer, 1992).[4]

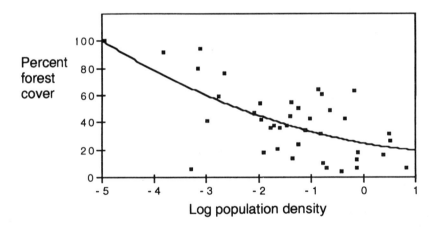

Source: Grainger (1986, 1992)

Figure 4.2 *Percentage national forest cover versus the logarithm of population density for countries in the humid tropics, 1980.*

Various factors threaten the move to a new equilibrium, however. Forest cover could continue to fall if farm yields do not rise rapidly enough, farming practices on cleared sites are unsustainable and lead to further clearance, and forest protection stays inadequate. Eventually, though, as in the UK, some lower limit should be

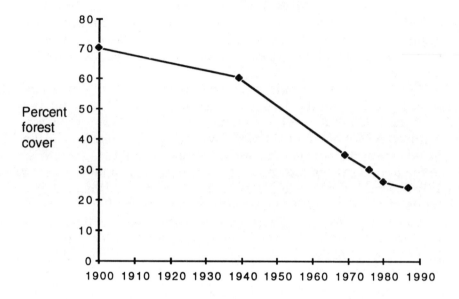

Source: based on data from Kummer (1992)

Figure 4.3 *Long-term trends in percentage national forest cover in the Philippines, 1900–1990*

reached. At the very least, some patches of forest will escape clearance since they are inaccessible or inside well protected national parks. But long before this the country will face mounting difficulties in meeting its own wood needs, pressures will grow to conserve forests for ecological reasons, and so economic and political forces should come into play to force private entrepreneurs and/or government to protect remaining forest from clearance. So whether percentage national forest cover levels off at a relatively high value (as in the USA) or a lower one (as in the UK) will depend on the vigour of the agricultural sector and the effectiveness of forest protection and conservation.

Whichever scenario is followed the national land use pattern will not remain static after equilibrium is reached. Major changes will still occur within existing agricultural and forestry areas as practices become more productive and sustainable, but further forest loss is unlikely. Indeed, as happened in the USA, forest area could even increase if yields carry on rising and national food needs can be met from a smaller area of land, allowing natural forest to regenerate and timber plantations to be established on surplus land. Alexander Mather, of the University of Aberdeen, has called this upward trend in forest cover a 'national forest transition' (Mather, 1990). It marks not only an end to deforestation, but also a sign that forest policy and management have reached a mature stage in a country.

The impacts of deforestation and logging on indigenous peoples

The continuing exploitation of tropical rain forests is seriously affecting the indigenous tribal peoples who still live in them. When the forests go they and their cultures go too. Some might consider these predominantly hunter-gathering peoples to be among the most 'primitive' in the world today, but from another perspective their cultures are as rich as the forest ecosystems that are their home. Instead of using pills when they become ill they take natural medicines derived from plants. They do not have printed lists of the many rain forest plants but know them nevertheless. If they disappear, we shall not only lose a part of the diversity of humanity but also their vast forest knowledge – much of which we would like to have ourselves.

The island of New Guinea, to the north of Australia, has a proud tribal heritage. Its western half is now part of Indonesia and being increasingly colonized as part of the Indonesian government's Transmigration Programme. Though the main aim of the programme is to resettle people from overcrowded Java to the less crowded 'Outer Islands' of this vast island nation, it also spreads the country's dominant Javanese culture as a basis for forging greater national unity and this inevitably leads to conflict with old tribal practices. In the east of the island, most of the 700 different peoples of the now independent Papua New Guinea survived the British colonial period but have since been drawn into the modern world by logging companies wanting to turn their forests into wood chips. Fortunately their communal system of land tenure still remains and is recognized by the government, enabling them in many instances to resist such pressures, however much the government has promoted logging.

Things are different in Sarawak, one of the two Malaysian states in western Borneo. There the land rights of some indigenous peoples are not acknowledged by the government, which reserves the right to decide how lands, forests and minerals are to be exploited. The Penan people are only deemed to have established rights over land if they clear the forest and cultivate the land underneath. Since 1986 they have protested against the spread of logging in their traditional forests by blocking logging roads so that concessionaires cannot carry their logs away (Sterba, 1987). In 1989, 71 Penans were arrested for disrupting logging. They failed in their aims but did draw world attention to their plight, though the Malaysian Prime Minister dismissed the Penan as 'unfortunate people' who were being exploited by 'crusading' environmentalists who wanted the tribe kept as ill-fed and disease-ridden 'museum pieces' (Hoon, 1989b).

Nowhere is the conflict between indigenous peoples and modern civilization more evident than in Brazilian Amazonia. For while forest dwellers there have been accorded increasing protection by government over the course of this century, and especially in the last 30 years, the government continues to override any rights it has acknowledged in order to implement highway, hydroelectric and other projects that further the aims of national development, and the effectiveness of government protection against unauthorized incursions into tribal territories, e.g. by miners, is so poor it can take months, if not years, to act against them.

The Indian population of Brazilian Amazonia has fallen ever since the Portuguese colonized the region in the sixteenth century, the victims of introduced diseases and wholesale slaughter. Now only 200,000 are left – just one fifth of the number in 1900 (Figure. 4.4). The largest group are the Yanomami, with a population of about 9000 in the state of Roraima and another 14,000 over the border in Venezuela. Within a year of their territory in Brazil being invaded in 1973 by workers building the BR 210 highway almost all of the people in 13 villages along the initial 100 km of highway had died from diseases caught from the construction teams. In 1988 over 40,000 miners invaded Yanomami land in Roraima in search of tin and gold, killing Indians both with guns and the diseases they carried. The miners were not driven out by the government until two years later (Matthews, 1989; Rocha, 1990a).

These examples of the impacts of deforestation and logging on indigenous cultures could be replicated by countless others in Amazonia and elsewhere. But things are changing. In 1982 the World Bank announced a new policy by which it would refuse to fund any project that encroached on the territories of indigenous peoples without their full and voluntary consent (Goodland, 1982). In 1983 it enforced this policy by stopping payments for work on Highway BR 364 in Amazonia since the government of Brazil had not created the Indian reserves specified in the project agreement. Payments only resumed when the government complied.

The Colombian government decided in 1988 to give Indians full legal title to their lands and by the end of 1989 had 'handed over' more than 18 million ha – almost half the total area of Colombian

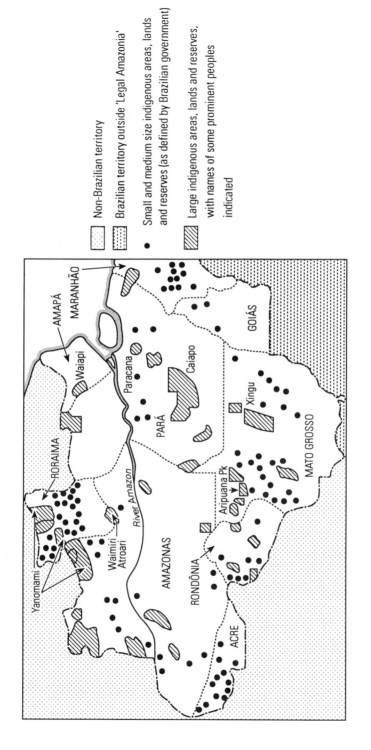

Figure 4.4 *Traditional lands of indigenous peoples in Brazilian Amazonia*

Amazonia – to 50 tribal groups comprising about 70,000 people. Recognizing the rights of indigenous tribal peoples in this way not only guarantees their survival also that of the tropical rain forests in which they live. The only doubt is whether such advances can spread sufficiently rapidly that enough forest dwellers will remain to do the protecting.

Forest policies and the spread of logging

Logging differs from deforestation not only in its impact on the forest but also in its underlying causes too. Deforestation occurs to meet demand for the land beneath the forest, logging to meet demand for the timber inside it. One of the saddest aspects of defor-estation is that in the rush to exploit forest land valuable timber is usually burnt after clearance rather than put to a productive use. In logging, on the other hand, trees are felled for their timber and a properly managed forest will continue to supply timber in the future. Another difference is that the spread of logging has been geographically far more selective than deforestation. The underlying cause of logging, demand for tropical hardwood, was discussed in Chapter 3. Here we look at why logging is often con-centrated in particular regions and countries and why forest management practices are currently less than optimal. The answers to these questions involve the consideration of economic factors and government policies, and allow us to identify institutional, as opposed to technical, constraints on sustainable forest management, understand historical trends in logging, and even predict possible future trends.

Understanding trends in logging

The tropical hardwood trade is centuries old, probably beginning quite early with the trading of a few valuable timbers, such as teak, on a regional scale, but it gained a new dimension after Europe's colonial and trading presence in the tropics grew from the sixteenth century onwards, and increased markedly in the nineteenth century with the spread of the Industrial Revolution in Europe and colonial power in Asia and Africa. But trade was still restricted to a few high quality decorative timbers, such as teak (from Asia), mahogany (Central America) and ebony and African mahogany (Africa), as well as woods used to produce dyes and tannins. Large quantities of teak were felled in India and Burma in the nineteenth century to build ships for the British Navy.

Only after the Second World War did the trade expand to include utility timbers with a wider range of uses. As tropical hardwood removals and exports grew so logging became increasingly concentrated in the forests of Southeast Asia, exploiting their vast reserves of light and medium-density timbers. Before long, 80 per cent of all exports came from that region. The Philippines was the first major post-war Asian exporter, in preference to Malaysia and Indonesia, though logging activity expanded there in the 1960s. Why was Southeast Asia preferred to Africa and Latin America, which had supplied significant shares of all exports in the immediate post-war years? Why were some countries preferred as export sources to others? Brazil and Zaire, for instance, contain about 40 per cent of all tropical moist forest but have played only a minor role in the tropical hardwood trade so far.

Many factors determine the spread of logging, including government policies, previous political and trading connections, and ease of transport to final markets. But the quality and quantity of timber in the forests have a major influence because they affect the economics of logging. Chapter 3 showed how only a few of the thousands of tropical tree species have marketable timbers. The commercial timber content of a hectare of forest is seldom above 30 per cent of its total timber volume and often much smaller. Southeast Asia's forests were so rich in valuable dipterocarp timbers that the commercial cut per hectare was up to ten times that in African and Latin American forests. The higher the commercial cut per hectare the lower the unit cost of logging, as the expense of building logging roads and trails – the dominant cost factor – is spread over a larger timber volume (Grainger, 1986). Other things being equal, forests with the lowest logging costs per cubic metre of wood would be preferred for harvesting. Other costs are involved, of course, particularly the cost of transport from the forest to the nearest mill or port, and from there to the final market. But most places in the Malesian archipelago (see Chapter 1) are relatively close to the sea, which means low internal transport costs, and Southeast Asia is near to Japan, so international transport costs to that major overseas market are also low.

The reason why one country is chosen in preference to others is similar: forests with the highest commercial cuts per hectare are more economic to log. This explains why first one country and then another has undergone logging booms. The Philippines was the first leading Asian exporter, owing to its commercial cut per hectare of 90 cubic metres. Only later did logging move to Malaysia (45–60 cubic metres) and Indonesia (45 cubic metres). In Africa, Nigeria (35 cubic metres) used to be a leading exporter. Then it declined and

Ivory Coast (25 cubic metres) became more prominent. The Philippines now exports very little tropical hardwood, while Nigeria has been a net importer of forest products for some time. The limited role of Brazil in the tropical hardwood trade until recently may be explained by its low commercial cut per hectare (5 cubic metres) – the lowest of any country – and high internal transport costs. The quality of timber in the forests of Brazilian Amazonia is generally quite low. Zaire's cut per hectare was greater than Brazil's (15 cubic metres) but it had high internal transport costs too. More evidence to support this view is that while commercial tropical hardwood reserves are concentrated in forests with low cuts per hectare, post-war exports came mainly from forests with high cuts per hectare (Figure 4.5). As logging spread from the richest forests to the slightly less rich so the average cut per hectare for all export removals declined, from 56 cubic metres in 1965 to 48 cubic metres in 1980 (Grainger, 1986).

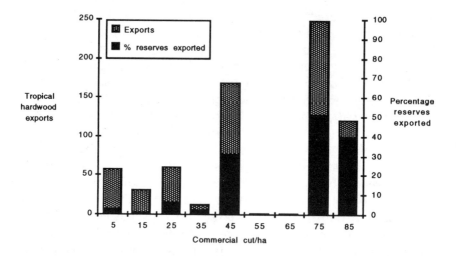

Source: Grainger (1986), calculated using FAO data

Figure 4.5   *Tropical hardwood exports 1965–79 (million cubic metres rwe) and percentage 1965 reserves exported by*

Within countries, logging is less spatially confined than deforestation because cuts per hectare are low compared with total timber volume and large areas need to be exploited every year. So instead of diffusing gradually over an area, as deforestation does, it could take place over a whole province simultaneously if concessions to all forests were allocated at the same time. As with deforestation, forests must be relatively accessible to be exploited, but logging is less restricted in this sense for if the cut per hectare is high enough concessionaires will build networks of roads to reach forests. Unfortunately, logging roads can promote the subsequent spread of deforestation – the process referred to earlier as 'logging-deforestation feedback'. Interestingly, the spread of logging in Brazilian Amazonia followed deforestation, rather than the other way round as was the case elsewhere, taking advantage of the better access that came after agricultural colonization and the building of the highway network.

Governments and sustainable management

However financially attractive logging might be governments still have to approve the start of large-scale logging in their countries, but the above evidence suggests their policies have usually been in step with economic considerations rather than the reverse, since the sequence of logging has tended to take place in order of decreasing national cuts per hectare. Logging is attractive because tropical hardwood is a valuable source of foreign currency needed for development, but if this is the case it is reasonable to ask why governments do not make sure that forest management is sustainable so this income can be perpetuated, instead of allowing tropical rain forests to be mined as if they were non-renewable resources.

The simplest answer is probably the best. Large areas of forests all too easily give the impression of being infinite and inexhaustible sources of timber until they are almost exhausted, and most governments can only think of maximizing short-term income. The fact that with good management the forests can be harvested again in 30–40 years does not have much meaning in this context. 'Old growth' forests being logged for the first time are in any case truly non-renewable resources, for they have grown unhindered for so long that the high density of valuable timber they contain is rarely regained under the kind of rotations common in normal commercial management. Moreover, the initial spread of logging in a country must necessarily form part of a wider colonization activity. Most governments broadly plan how the total area of land is to be used in future, and though some areas of forest are scheduled to remain for long-term timber production or conservation others will be designated for conversion to agriculture or other uses. It is preferable, of course, if valuable timber can be exploited before forest is cleared but often this does not happen. Even in areas of forest designated for timber production initial logging activities will be exploratory, providing the basis for sustainable management in future but not at the moment satisfying many of the criteria for sustainability listed in Chapter 3.

Forest policies: ideals versus reality

But what is the role of forest policies? Many governments do not actually publish their forest policies, and those that do rarely live up to the fine sounding words within them about the need for sustainable management and environmental protection. When policies are not published all we can do is to assume that what actually transpires in the forest conforms with the government's real policy, for otherwise it would take corrective action.

Ideally, government forest policies should ensure that initial logging will be followed by sustainable forest management by:

1. deciding the ownership of forests, and granting rights to harvest them;
2. regulating how forests are managed and harvested;
3. determining the rates of depletion and replenishment of timber reserves;
4. determining the prices at which timber from national forests is sold and the charges levied on those who log the forests;
5. giving financial support to forestry activities on private lands, as appropriate;
6. protecting forests from disease, fire, and illegal encroachment;
7. integrating forest policy with other policy objectives, such as creating employment and conserving the environment.

In the early stages of tropical rain forest exploitation governments give priority to only a few of these policy responsibilities and neglect the others. They allocate logging concessions and specify the selective logging system to be used, but give little attention to monitoring and enforcement so target rates of timber production are exceeded and the status of forest regeneration for the second harvest is uncertain. They give priority to raising revenue too but often do not set the various fees and taxes paid by concessionaires as high as the market could stand so their share of the returns from logging is not as high as it could be (see Chapter 3) (Repetto and Gillis, 1988). (Claims that this is the reason for the low cuts per hectare in tropical rain forests are misguided.) Finally, they do not pay much attention to forest protection and the wider environmental effects of forest use, and in many, but not all, countries the emphasis is simply on timber exploitation rather than building a vertically integrated forest industry that contributes to long-term national economic development.

The divergence of current logging practices from the ideals laid out in forest policies therefore seems to represent a policy failure of the first magnitude. Two foresters with long experience in the tropics have tried to explain this. According to John Palmer, of the Oxford Forestry Institute, the policies proposed by foresters are technically sound but practically naive in the wider context. They may contain some fine ideals but, because of the general political inexperience of foresters, they lack any relevance to the very pragmatic requirements of governments for cash (Palmer,1989). The late Jack Westoby, a former head of the FAO Forestry Department, took a less charitable view. He spoke scathingly of forest policies being hatched between top foresters and politicians so that powerful commercial interests are accommodated and the policies are vague enough to be changeable as conditions warrant (Westoby, 1987). Politicians and other prominent individuals often exert considerable patronage in allocating concessions to friends, relations and others. When the government of Thailand announced a ban on all logging in 1988, due to forest depletion and environmental concerns, senior military officers are said to have arranged new concessions for Thai companies in Burma, and Thailand's imports of Burmese teak were soon four times the level of its domestic removals before the logging ban (Matthews, 1990; Lintner, 1992).

To these perspectives we must add another. However good or bad policies may be on paper their ultimate impact depends upon how well they are implemented. The limitations on policy implementation owing to lack of personnel and organizational capacity have been referred to earlier. But central policy directives can also be hampered by different policies at local level. For example, Malaysia has a federal constitution, and responsibility for forest management lies with the states, such as Sarawak and Sabah, rather than with the federal government in Kuala Lumpur. Even in Indonesia, the prerogatives of provincial administrations present an effective filter between what is desired in Jakarta and what happens in the forest.

Is logging dominated by foreign companies?

Large transnational corporations are often blamed for dominating tropical rain forest logging. But while overseas interests, represented by companies based in the USA, Japan, South Korea and European countries, are significant this is an overgeneralization. For example, in the 1970s, the boom period for logging, 26 major US-based trans-nationals with forest products interests, such as Weyerhaeuser and Georgia Pacific, were involved in only 15 instances of forest management in six countries (Costa Rica, Brazil, Colombia, Indonesia, Malaysia and the Philippines) and four of these were pulpwood plantations (Bethel et al, 1982). A wider study concluded that in the late 1970s foreign firms were involved, either individually or in joint ventures with local firms, in 25–35 per cent of all logging in Asia-Pacific. Involvement was twice as great in African logging but minimal in Latin America, but while French firms dominate logging and wood processing in Francophone countries such as Gabon, the Congo and Cameroon, Africa only accounts for a small fraction of all tropical hardwood exports and removals and so worldwide their impact is relatively minor. Foreign involvement should not be equated automatically with large transnational corporations, however. Of all overseas investment in Indonesian logging between 1967 and 1979, for example, over half came from other Asian developing countries (Don Scott-Kemmis, Personal communication).

Plantations versus natural forest management

Since the 1960s tropical forest policy has been dominated by the plantation ethos, so establishing plantations has been given priority over managing natural forests. This arose in part from the relative failures of experiments in tropical silviculture in the first half of the century (see Chapter 7). Plantations have an enormous potential, since current annual demand for tropical hardwood could, according to a rough estimate by this author, be produced sustainably on just 28 million ha of plantations, i.e. seven times the annual rate of tropical rain forest logging (Grainger, 1993). Concentrating timber production in this way would also avoid extensive logging, the associated forest degradation, and its catalysis of subsequent deforestation by farmers. But previous planting rates have been far too low to realize this potential in the foreseeable future and most new plantations in recent decades have been designed to produce pulpwood and fuelwood, rather than high grade hardwood such as that from tropical rain forests. High grade hardwood plantations account for only 18 per cent of total tropical timber plantation area and less than 1 per cent of all tropical hardwood production. That share is unlikely to rise above 5 per cent in the next 40 years (see Chapters 7 and 9) (Grainger, 1986, 1988a) and higher tropical hardwood prices – which are indeed likely as logging shifts from low cost South-east Asia to higher cost Latin America and Africa – will be needed to encourage a rise in the rate of planting. So while a strong plantation programme is still important a major shift in forest policy is essential to give a new emphasis to natural forest management.

Values and scarcity

There has been a growing appreciation recently that tropical rain forest is economically undervalued relative to its total biological wealth and the environmental services it provides. It is true that many of the benefits of tropical rain forest do not have a market price, and this is so for renewable resources all over the world (see Chapter 7) but what of the goods that are marketed – timber and minor forest products? Are they undervalued? Should prices be increased to put long-term forest management on a better financial footing? Minor forest product harvesting is discussed in Chapters 6 and 7. Here we comment briefly on timber.

Charting trends in the price of 'tropical hardwood' is difficult since prices vary greatly by product, species, quality and country of origin, and consistent, reliable price data are only available for some species and countries. Here average unit export values based on FAO data are used to show general trends (see Grainger (1986, 1993) for a more detailed analysis). Tropical hardwood prices are set on the world market, reflecting competition between various suppliers and substitute woods. Those who claim the price of tropical hardwood is too low might imagine it is still mainly a luxury timber, whereas a large proportion is now used for utility purposes. Consequently, trends in average real tropical hardwood log prices over the last 30 years followed those for all hardwood logs (i.e. including those from temperate forests) and all softwood logs (Figure 4.6). If tropical hardwood prices were artificially raised, either by producing countries forming a cartel for collective commercial gain or importing countries imposing surcharges to try to 'save the rain forests', it would probably just lead to greater demand for temperate substitutes rather than higher income for tropical countries.

Is tropical hardwood becoming scarcer as timber reserves are depleted by logging and deforestation? We would expect any increasing economic scarcity to be signalled by rising prices but though the tropical hardwood log price increased by three and a half times between 1965 and 1987 it remained fairly constant in real terms (Figure 4.6), in accord with the FAO price index for all forest products worldwide. A detailed analysis by this author showed no

Controlling tropical deforestation

A. Prices of tropical hardwood logs and all hardwood logs

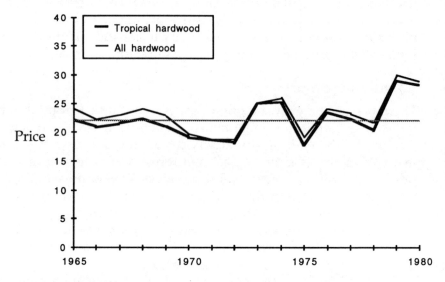

B. Prices of tropical hardwood logs and all softwood logs

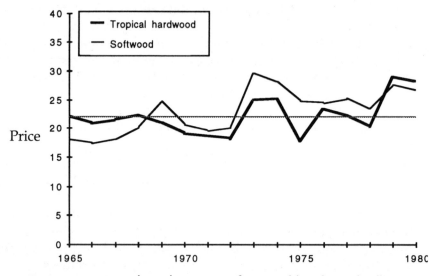

Figure 4.6 *Trends in the prices of tropical hardwood, all hardwood and all softwood logs (1965 $US per cubic metre)*

indication by prices of a growing long-term scarcity of tropical hardwood either on a regional or national basis, even in the Philippines where forests and timber reserves have been seriously depleted. The long-term real price of sawn tropical hardwood has also stayed about the same but that of plywood has fallen sharply with rising production. Average unit logging costs were fairly constant over the same period because of the dominance of Southeast Asian producing countries, but as logging shifts to higher cost Latin American and African reserves in the next twenty years an upward pressure on prices would be expected (Grainger, 1986).

Developing forest industries

Tropical hardwood is one of a number of primary commodities on which tropical countries rely for export earnings. Just under a half of all tropical hardwood is still exported as logs rather than sawn-wood, veneers and plywood. FAO and other international agencies have consistently advised the governments of tropical countries to increase domestic processing capacity in order to develop a verti-cally integrated forest industry and raise their share of the value added to tropical hardwood by processing. If timber is exported as logs this value goes to the importing country instead. Japan, Europe and the USA tried to perpetuate the traditional pattern by imposing tariffs of up to 10 per cent on sawnwood imports and up to 20 per cent on plywood (Takeuchi, 1981) but these barriers were relaxed during the 1980s. Domestic processing can also increase the number of species extracted from the forest since local mill owners can be much more flexible than foreign timber traders when choosing species.

Despite the benefits of pre-export processing many producing countries have so far failed to expand processing capacity to any great extent. One reason for this has been a shortage of investment capital, allowing the forest-poor Other Asian Processing Nations (South Korea, Taiwan, Hong Kong and Singapore) to become the leading processors of Asian tropical hardwood in the 1970s. But a strong domestic market is also necessary, so that sawmills and plymills do not have to risk their entire investment on the vagaries of the world market. Indonesia's ability to phase out log exports in the early 1980s, build 78 new plymills in just six years and switch to exporting processed wood had a lot to do with its vast domestic market, something lacking in the other leading producing country, Malaysia.

Indonesia succeeded in establishing a major forest industry, and raised its share of all tropical plywood exports from 5 per cent in 1980 to 59 per cent in 1987, but at a heavy cost to the government due to lost revenue from log exports and the subsidies it had to give

(Fitzgerald, 1986). As a significant proportion of all domestically processed wood is consumed at home (regional averages range from 60–90 per cent for sawnwood and 20–75 per cent for plywood) an exact replacement of log exports by the same roundwood equivalent volume of products is unlikely, and this lowers the potential for increasing export earnings (Grainger, 1986). Processing efficiencies will also probably be low as new forest industries develop so this can, as happened in Indonesia, lead to an increase in log production and so exacerbate, rather than control, the spread of logging.

Deforestation, logging and development

Deforestation is an inevitable part of a country's social and economic development, occurring in response to increases in population and economic activity and promoted also by government policy. These are the major underlying causes of deforestation, although poverty, inequality and unsustainable agriculture also play important roles and present land uses owe a lot to historical influences, including those derived from colonial rule. Whether or not large-scale logging takes place as well also depends on governments but is very much determined by international market conditions and the cost effectiveness of logging in the country concerned.

In the early years of forest exploitation in a country, when forest protection is generally given a low priority, agriculture is largely free to expand at the expense of forests. Lack of sustainability of forest management results from a failure of government forest policies and institutions and the need to maximize short-term income both for national development purposes and sheer entrepreneurial greed. However, complete deforestation is not inevitable if agricultural productivity and sustainability can expand quickly enough, forests are protected for long-term timber production and ecosystem conservation, and governments do not subsidize unsustainable or unproductive agriculture or construct roads unwisely through forested areas. These factors should ensure that the decline in forest cover eventually levels off, the ultimate limit being represented, if necessary, by government intervention to secure a minimum level of forest cover to supply domestic wood needs and conserve the environment.

Logging is likely to continue in the tropical rain forests, the focus of activity spreading more and more from Southeast Asia to Latin America and then Central Africa over the next decade. But whether these forests fulfil their potential as long-term suppliers of hardwoods to the world market depends on whether forest management becomes more sustainable, and in the short-term this means giving greater priority to protecting forests from deforestation and ensuring that logging practices are in accord with government guidelines. Since most tropical hardwood is going to come from these forests rather than from plantations in the foreseeable future, governments must place greater emphasis in their forest policies on natural forest management and so correct the present bias towards plantations.

The implications for programmes to control deforestation and improve forest management are that action is needed in the agriculture sector as well as the forest sector, and that policies must be changed as well as land management techniques. Attention should be given not only to forest and agricultural policies but also to social and economic policies which initially may not seem to be directly connected with changes in land use and forest cover. Some suggestions for policy changes are made in Chapter 8.

The scale of deforestation

The deforestation of tropical rain forest is a major global environmental problem, occurring in more than sixty countries simultaneously. If it continues unchanged at current rates very little tropical rain forest will remain in a hundred years' time, although Chapter 4 urged caution in projecting current rates into the future. But how big a problem is it really? This is not an academic question, for the size of the annual deforestation rate critically affects the degree of public and political concern about the problem: the higher the estimate, the more likely policy makers are to take action to control deforestation. The degree of uncertainty attached to estimates also has an influence. Sadly, until 1992 we did not know either the total area or deforestation rate for tropical rain forest, only the figures for tropical moist forest (see Chapter 1), and these were very inaccurate, making it difficult also to assess the long-term impacts of deforestation on global climate, biological diversity and tropical hardwood supplies. This chapter reviews the estimates made over the last 20 years of the deforestation rate for tropical moist forest, examines their reliability, and suggests how to obtain better estimates in the future.

Where do estimates come from?

The range of available estimates of deforestation rates can be confusing to the person who simply wants the most reliable and up-to-date figure. Some are just official government statistics; others come from the UN Food and Agriculture Organization (FAO) or independent specialists such as Norman Myers; and there are also estimates from various other sources, though most of those are just derived from the primary estimates of FAO and Myers and are not genuine alternatives to them.

Global and national estimates can vary, often quite markedly, as we shall see below. There are a number of reasons for this, the two most elementary being differences in the way deforestation is defined and in the types of tropical forest included in the estimate. An estimate from FAO, for example, which refers strictly to deforestation as it is defined in this book, i.e. forest clearance (see Chapter 1), will be lower than one that includes both the clearance and logging of primary forest and the clearance of forest fallow. The rate of deforestation for tropical moist forest will be greater than that for tropical rain forest, but lower than the rate for all tropical closed forest (both moist and dry). The rate for all tropical forest, which includes both closed forest and open forest (see Chapter 1), could be half as much again.

These are relatively easy sources of variation to deal with. It is much more difficult to evaluate the reliability of estimates that really do refer to the same thing. The most obvious division is between estimates based on measurements and those relying on the subjective judgement of experts or the authority of governments. For many years official figures were the only primary source of data. So FAO and independent experts had to decide whether to accept them at face value or use their own judgement as to the most likely rate, based on both official and unofficial data, and perhaps making use of models as well. While such estimates are made with the best of intentions they are highly subjective and their reliability is hard to assess.

Since remote sensing satellites, such as the US Landsat, have continuously monitored the Earth's surface for more than 20 years, many people might imagine that we have taken full advantage of all the data collected by satellites to monitor tropical deforestation, but they would be wrong. For the potential to use the full range of remote sensing techniques in this area, including aerial photography and side-looking airborne radar as well as satellites, has not yet been realized (Lillesand and Kiefer, 1987; Grainger, 1984). If an estimate is stated to be based on remote sensing measurements and the particular technique is quoted then some idea of the accuracy of the estimate may be gained, but if the technique is not mentioned we must assume that it depends on expert judgement only. It is also important to distinguish between estimates of the actual rates of deforestation over a previous period and forecasts of possible rates in the future – both kinds of figures are released by FAO and they should not be confused.

Even if an estimate is based on measurements its accuracy is still not guaranteed. Measurements made using medium to high resolution sensors on the Landsat and SPOT satellites, for instance, are more reliable than those made by low resolution AVHRR sensors on NOAA weather satellites. A Landsat MSS sensor, for example, has a resolution of 79 m, which means effectively that the sensor treats all solar radiation reflected from objects within a 0.62 ha area on the ground as one response. The resolution of AVHRR sensors, on the other hand, is much lower, either 1 km or 4 km, so while the MSS sensor can distinguish between forest and non-forested areas of a few hectares in size, the AVHRR sensor will miss many small clearings. The use of photographic versions of images in deforestation monitoring will also overlook small clearings that computer-based analysis of the original digital versions of the images will pick up.

Sometimes only part of a country is surveyed, and the resulting estimate will be inaccurate if the area studied was unrepresentative of the whole country and statistically-based sampling was not used. Estimates of deforestation rates are made by measuring the change in forest cover over a certain period. The reliability of an estimate would be reduced if the forest area estimates at the start and finish of this period were each based on measurements spread over a number of years. This could arise if only a few images had been collected in one year and they had been of poor quality because of heavy cloud cover, which is still a major obstacle to remote sensing in the humid tropics but one that could be removed by the new generation of radar satellites. Alternatively, a remote sensing survey in one year could be compared with an older forest map, but if the latter was based on poor data – and often map sources are not mentioned – then the estimated deforestation rate will again be of questionable value. This all shows the need for care when evaluating estimates of deforestation rates, and that we should not assume that the most recent estimate is necessarily also the best.

Estimates of deforestation rates

Uncertainty about the deforestation rates for tropical moist forest should be seen in the light of the variations between the different estimates of total forest area reviewed in Chapter 1. Only about 60 per cent of the original climax forest cover – the area that should have existed according to climatic conditions before any human disturbance – is thought to have remained in the mid-1970s (Sommer, 1976). But there is no way of knowing what the original area was and estimates of historical trends in forest area, even in quite recent times, are very sketchy. If there were still 1,081 million ha of tropical moist forest left in 1980, as proposed in one estimate (Grainger, 1983), it would, if all collected together, fill a land area 10 per cent larger than the USA. This estimate was actually greater than those published in the 1970s, and although since then Norman Myers has said that the area at the end of the 1980s had fallen to 800 million ha, in the last few years FAO released an interim estimate of 1,282 million ha for the area in 1990, with a more authoritative figure expected in early 1993 (Table 5.1).

Table 5.1   Estimates of tropical moist forest area (million ha)

Source	Date	Area
Persson (1974)[a]	1970s	979
Sommer (1976)	1970s	935
Myers (1980)[a]	1970s	972
Grainger (1983)	1980	1,081
Myers (1989)	1989	800
FAO (1990)[b]	1990	1,282

a: Derived from source data by Grainger (1984)
b: Revised for the UN Conference on Environment and Development in 1992

Sommer's 1976 estimate

The earliest estimates of tropical deforestation rates date from the 1970s, when the commonly accepted rate in the humid tropics was between 11 and 15 million ha per annum. This was often simplified to 15 million ha per annum and on one occasion, in an address by the Director General of FAO to the Eighth World Forestry Congress in 1978, to 30 ha per minute (Table 5.2). The latter was equivalent to 16 million ha per annum but conveniently translated to an area of forest the size of an British football pitch lost every second (Grainger, 1980a). All these rates came from an original estimate by Adrian Sommer for FAO of 'at least 11 million ha per annum', made by assembling estimates of annual deforestation rates for just 13 countries[1], calculating the average percentage deforestation rate and extrapolating that to the total area of tropical moist forest. The sources of the estimates were not mentioned and the 13 countries contained only 16 per cent of all tropical moist forest (Sommer, 1976).

Myers' 1980 estimate

In 1980 the US National Academy of Sciences published a report by Norman Myers that outlined the scale and causes of human impacts on tropical moist forest in the leading countries in the humid tropics. He used the term 'conversion' to refer to all impacts 'from marginal

Table 5.2   *Estimates of rates of deforestation in the humid tropics (million ha per annum)*

Source	Date	Period	Total
Sommer[a]	1976	1970s	11–15
Myers[b]	1980	1970s	7.5–20
Grainger[c]	1983	1976–80	6.1
Myers	1989	late 1980s	14.2
FAO[d]	1990	1981–90	16.8
FAO[e]	1992	1981–90	12.2

a: 15 commonly quoted
b: 7.5 a later revision
c: cf 7.3 for all tropics
d: interim estimate
e: revised interim estimate presented to UN Conference on Environment and Development

modification to fundamental transformation' but only included estimates of deforestation rates for 13 countries[2]. Extrapolating the sum of these rates, 7.8 million ha per annum, to the whole humid tropics would have given an annual deforestation rate of 11 million ha (Grainger, 1984) but Myers himself did not attempt to make a rigorous estimate of the total rate of deforestation or conversion using his national estimates, merely commenting that it was possible that Sommer's estimate for FAO was significantly low, and that a better estimate of the annual rate of forest disappearance could be over 21 million ha (Myers, 1980). This seemed to refer only to deforestation but it was later revised to a loss of 7.5 million ha of closed forest and 14.5 million ha of forest fallow (Melillo et al, 1985), making it roughly compatible with the FAO/UNEP estimate, mentioned below.

The 1980 FAO/UNEP Tropical Forest Resource Assessment

In the following year FAO and the UN Environment Programme (UNEP) published a huge 1,500 page report on the state of the tropical forests. The total area of all tropical closed forest (i.e. both moist and dry) in 1980 was estimated as 1,201 million ha and the annual deforestation rate between 1976 and 1980 as 7.3 million ha (Lanly, 1981). Deforestation referred only to forest clearance, as in this book. A later summary report *projected* the average annual deforestation rate for open woodlands in dry areas for 1981–5 as 3.8 million ha (Table 5.3) and the average annual area of all tropical closed forest logged in the same period as 4.4 million ha (Table 5.5) (only regional figures were included in both cases) (Lanly, 1982).

Table 5.3 *Annual rates of deforestation for tropical closed forest and tropical moist closed forest (1976–80), tropical dry open woodland (projections 1981–85), and all tropical forest (million ha per annum)*

	Closed forest All[a]	Moist[b]	Open woodland[c]	All tropical forest[c]
Africa	1.3	1.2	2.3	3.6
Asia-Pacific	1.8	1.6	0.2	2.0
Latin America	4.1	3.3	1.3	5.4
Total	7.3*	6.1	3.8	11.1

NB: *Totals are not the sum of regional figures due to rounding. All tropical forest is the sum of all closed forest and open woodland.
Sources: a: Lanly (1981); b: Grainger (1983), based on Lanly (1981); c: Lanly (1982)

This report was a milestone. For each of 76 countries it estimated the national forest area and deforestation rate using a common system of forest resource classification, corrected all estimates of forest areas and deforestation rates to the same year (1980) and period (1976–80) respectively, gave forest areas for most tropical countries for the first time since the 1965 FAO World Forest Inventory and most national deforestation rates for the first time ever, and also made projections of national deforestation rates in closed forests for 1981–5[3]. It listed in many cases the sources of data for estimates of forest areas and deforestation rates and the measurement techniques, if known. But though remotely sensed measurements were a main data source for the forest area estimates for a quite a number of countries, only a few estimates of deforestation rates were derived in this way.

Grainger's 1983 estimate

The FAO/UNEP deforestation data for all tropical closed forest were adapted by this author to give a secondary estimate for tropical moist forest only, omitting the deforestation in tropical dry closed forest. Estimates of national deforestation rates were listed for 55 of the 65 countries that Sommer (1976) said constituted the humid tropics (Tables 5.4 and 5.5). They accounted for 99.5 per cent of the total area of tropical moist forest, but as reliable estimates of rates were unavailable for another six countries the total rate for the 55 countries, 6.1 million ha per annum, was taken as the estimate for all tropical moist forest. The annual loss of forest, equivalent to the size of Sri Lanka, represented a reduction of 0.56 per cent in the total forest area, estimated in the same report as 1,081 million ha. Over half of all deforestation took place in Latin America and a quarter in Asia-Pacific (Grainger, 1983, 1984).

Table 5.4   *Estimates of regional rates of deforestation in the humid tropics (million ha per annum)*

	Grainger (1983) 1976–80	Myers (1989) late 1980s	FAO (1992)[b] 1981–90
Africa	1.2	1.6	3.0
Asia-Pacific	1.6	4.6	2.6
Latin America	3.3	7.7	6.6
Total	6.1	14.2[a]	12.2

a: Regional figures are totals of individual countries listed by Myers. Total was adjusted by him for the whole biome-type.
b: FAO interim estimate of 16.9 million ha per annum for all tropical forest disaggregated by FAO for UN Conference on Environment and Development in 1992.

Table 5.5   *Rates of logging in the tropical forests (million ha per annum)*

	Estimates[a] for tropical moist forest 1976–80	Projections[b] for tropical closed forest 1981–5
Africa	0.8	0.6
Asia-Pacific	1.3	1.8
Latin America	1.6	2.0
Total	3.7	4.4

Sources: a: Grainger (1986), b: Lanly (1982)

This estimate was much lower than Sommer's, largely because FAO/UNEP's estimate for tropical closed forest was lower, but it seemed more reliable due to FAO/UNEP's more systematic approach. Nevertheless, its accuracy was still open to question, as only six national deforestation rates had been estimated in whole or in part based on remote sensing surveys. For another seven countries local estimates of unverifiable accuracy had been used, and for the remaining countries estimates had been made by FAO staff using best available information.

In addition to the 6.1 million ha of tropical moist forest deforested each year this author calculated that on average a further 3.7 million ha per annum was logged between 1976 and 1980 (Grainger, 1986). Latin America accounted for two fifths of this and Asia-Pacific for a third (Table 5.5). This compares with FAO's projection that 4.4 million ha of all tropical closed forest would be logged each year between 1981 and 1985 (Lanly, 1982).

Myers' 1989 estimate

Norman Myers published a second report on tropical moist forests in 1989 that paid more attention to estimating forest areas and deforestation rates than his earlier study, listing these for 24 countries individually and ten other countries in two regional groups, Central America and the Guianas (Guyana, French Guiana and Surinam) (Tables 5.4 and 5.8). The total deforestation rate for the 34 countries (which contained 97 per cent of all tropical moist forest) was 13.8 million ha per annum, which Myers extrapolated to 14.2 million ha per annum for all tropical moist forest. Just over half of the total deforestation rate took place in Latin America and Asia-Pacific's share was about one third (Myers, 1989).

Comparing this with the FAO/UNEP estimate (rather than his own earlier figure) Myers claimed that the global deforestation rate had doubled during the 1980s. There were increases for 26 countries, decreases for six countries and no change for one country[4]. The average percentage deforestation rate had gone up from 0.6 per cent to 1.8 per cent – remember Myers had also given a much lower estimate of total forest area of 800 million ha (see Chapter 1). But how reliable were his figures? Had deforestation rates really doubled? Economic conditions had certainly changed over the decade, with the world recession and Third World debt crisis, but on the other hand the expansion of cattle ranching in Amazonia and the Indonesian Transmigration Programme had both peaked by then. The reliability of the estimates is difficult to assess, for while Myers gave copious references he usually did not tell the reader which measurement techniques had been used, if any, and to what extent national estimates depended on the expert judgement of local experts or Myers himself.

For some countries the data sources listed allow a critical evaluation to be made. For example, the rate for Brazilian Amazonia was given as a conservative 5 million ha per annum. This happens to be a rough mean of two conflicting estimates: 1.7 million ha per annum for 1978 to 1988 made by one group at the Brazilian National Space

Agency (INPE)(quoted in Da Cunha, 1989) and 8.1 million ha per annum for 1987 made by another INPE group (Setzer et al, 1988). The higher figure was later disowned by INPE as an exaggeration since it was made by using imagery from the low resolution AVHRR sensor on a NOAA satellite to measure not the change in forest area but the number of fires in part of Amazonia (Marcio Barboso, Personal communication, 1990). The lower figure was a slight underestimate but closer to the likely rate of about 2.1 million ha per annum (see below) (Fearnside et al, 1990). Myers' estimates for Indonesia, Malaysia and the Philippines were all at least twice the FAO/UNEP 1976–80 rates but referred to both deforestation and selective logging. Correcting for the logging rates and adjusting the Brazilian figure (to 2.1 million ha per annum) would cut the deforestation rate for all tropical moist forest to about 10.3 million ha per annum. A closer appraisal could reduce it even more.

Recent FAO estimates

FAO was still working on a sequel to its 1980 report as this book was being completed. Estimates of tropical forest areas in 1990 and deforestation rates in the 1980s – both based on satellite image data – were due for publication in early 1993. An interim estimate of 16.8 million ha per annum for the average deforestation rate for tropical moist forest during the period 1981–90, issued in 1990, relied heavily on extrapolation and modelling rather than measurements, so its reliability was doubtful (FAO, 1990). It was of the same order as Myers' 1989 estimate, so reservations about that must apply to this one too.

In the following year FAO stated that this figure for tropical moist forest alone had been too high, and issued a revised interim estimate of 16.9 million ha per annum for the deforestation rate for *all* tropical forest, i.e. closed and open forest in the humid and dry tropics (Table 5.4) (FAO, 1991). It covered 87 countries, used a similar methodology to the 1990 estimate and should be compared with the earlier FAO/UNEP estimate of 11.1 million ha per annum for all tropical forest (Table 5.3). In 1992 FAO presented a disaggregated version to the UN Conference on Environment and Development, estimating the deforestation rate of tropical moist forest as 12.2 million ha per annum, which included a rate of 4.9 million ha per annum for tropical moist forest. As this was still an interim estimate it is probably best to wait until FAO's final report before making a definitive judgement on deforestation rates in the 1980s.

Choosing the best estimates

This section has examined the major estimates of deforestation rates in the humid tropics over the last twenty years and found them wanting. Using the criteria suggested at the start of the chapter we might conclude on balance that the 6.1 million ha per annum estimate for the late 1970s, based on FAO/UNEP data, was the most reliable of all the estimates, but still with reservations owing to the lack of measurements involved. The Myers (1989) and various FAO (1991) estimates are rather unsatisfactory, and an estimate wholly based on remote sensing measurements is urgently awaited. Finally, did the annual deforestation rate really vary so much between 1970 and 1990? Instead of falling from 15 million ha to 6.1 million ha and rising to 12.2–14.2 million ha or even higher, if the first rate was close to 11 million ha, the second an underestimate, and the last an over-estimate then perhaps, given the large errors involved, it did not. It would be nice to find out for certain by analysing the satellite imagery for this period – while it is still available.

Even less is known about historical trends in tropical moist forest area. The best estimate for the first decade or two of this century, made by this author using the earliest global forest survey, published in 1923 by Zon and Spaarhawk, was 1,120 million ha, well inside the limits of error of the estimate for 1980. Adding to the latter the 115.6 million ha of forest in the humid tropics thought to have been cleared for agriculture between 1860 and 1978 (Williams, 1990; Revelle, 1984; Richards et al, 1983; Richards, 1986) gives a mid-nineteenth century area of 1,197 million ha. For what they are worth, these suggest average deforestation rates of 1.3 million ha per annum between 1860 and 1920 and 0.65 million ha per annum between 1920 and 1980, but they are not worth very much.

Deforestation at regional and national level

Global deforestation rates are made up of a diverse collection of trends in individual tropical regions and countries. Three sub-regions which stand out are West and East Africa and South Asia, for according to Adrian Sommer by the mid-1970s they had lost 72 per cent, 72 per cent and 64 per cent of their original climax forest cover respectively[5] (Table 5.6). Elsewhere the loss was only about 40 per cent, close to the average for all tropical moist forest (Sommer, 1976). Within each region the size of the deforestation problem varies greatly. This section is a brief survey of the leading countries in the humid tropics, based on a longer analysis in Grainger (1993), using mainly the deforestation rates estimated by Grainger (1983) for 1976–80 – based on FAO/UNEP data (Lanly, 1981) – and by Myers (1989) for the late 1980s.

Table 5.6   *Proposed reductions in tropical moist forest area from climax area by 1976 (million ha)*

	Climax area	Current area	Regression	Percentage loss
Africa	362	175	187	52
East	25	7	18	72
West	68	19	49	72
Central	269	149	120	45
Asia-Pacific	435	254	181	42
South	85	31	54	64
South-east	302	187	115	38
Pacific	48	36	12	25
Latin America	803	506	297	37
Central	53	34	19	36
South	750	472	278	37

Source: Sommer (1976)

Africa

Africa had 205 million ha of tropical moist forest in 1980, a fifth of the world total, and was losing it at a rate of 0.6 per cent per annum (Grainger, 1983). Four fifths of the forest was in just four countries in Central Africa: Cameroon, Congo, Gabon and Zaire (Tables 1.5 and 1.6). The overall loss of forest in this sub-region relative to the climax area was, at 40 per cent, close to the global average, according to Sommer (1976). All four countries have only moderate population densities, their annual percentage deforestation rates are close to the regional average (Table 5.8), and so deforestation does not represent much of a threat. Zaire contains about a tenth of all tropical moist forest but there have been no reliable surveys in recent decades so the estimated 1980 forest area of 105.7 million ha and 1976–80 deforestation rate of 165,000 ha per annum (Table 5.7) should be treated with caution.

The situation is different in West Africa, which may have already lost over 70 per cent of its climax forest and accounts for over half of the total African deforestation rate. Deforestation has been extensive all over the region, but has reached critical levels in Ivory Coast and Nigeria. They accounted for 57 per cent of all tropical moist forest in West Africa in 1980 but were losing 7.0 per cent and 4.8 per cent of

Table 5.7. *Two estimates of national rates of deforestation of tropical moist forest, 1976–80 and late 1980s (thousand ha per annum)*

	Grainger 1976–80	Myers late 1980s
Africa		
Cameroon	80	200
Central African Republic	5	na
Congo	na	70
Equatorial Guinea	15	na
Gabon	27	60
Ghana	15	na
Guinea	na	na
Guinea-Bissau	15	na
Ivory Coast	310	250
Liberia	41	na
Madagascar	165	200
Nigeria	285	400
Sierra Leone	6	na
Uganda	10	na
Zaire	165	400
Asia-Pacific		
Brunei	7	na
Burma	89	800
India	na	400
Indonesia	550	1200
Kampuchea	15	50
Laos	120	100
Malaysia	240	480
Papua New Guinea	21	350
Philippines	100	270
Thailand	325	600
Vietnam	60	350
Latin America		
Belize	9	b
Bolivia	65	150
Brazil	1360	5000
Colombia	800	650
Costa Rica	60	b
Cuba	2	na
Dominican Republic	2	na
Ecuador	300	300
El Salvador	4	b
French Guiana	64	a
Guatemala	1	b

Controlling tropical deforestation

	Grainger 1976–80	Myers late 1980s
Honduras	53	b
Nicaragua	97	b
Panama	31	b
Peru	160	350
Surinam	3	a
Venezuela	125	150

a: Total deforestation rate for French Guiana, Guyana and Surinam given as 50,000 ha per annum (cf 70,000 ha per annum in 1976–80)
b:Total deforestation rate for Belize, El Salvador, Honduras, Nicaragua, Costa Rica and Panama given as 300,000 ha per annum (cf 255,00 per annum in 1980).
na indicates data not available.
Sources: Grainger (1983), Myers (1989)

their forest every year respectively in the late 1970s. Norman Myers (1989) thought the Ivory Coast deforestation rate fell in the 1980s and the Nigerian rate rose. Nigeria is a large country, containing one in every six Africans, and was once a leading tropical hardwood exporter but after extensive agricultural development only 6 million ha of forest were left in 1980. Ivory Coast is also prosperous but the spread of agriculture cut its forest area by 60 per cent between 1950 and 1980, when only 4.5 million ha of tropical moist forest remained, and this may have declined to 1.6 million ha by 1989, according to Myers (1989). Even in 1980 it was Africa's leading log and sawnwood exporter but over the next seven years removals fell by 50 per cent and log exports by 80 per cent.

Madagascar is remarkable for its large number of endemic species: some 8,000 plant species are found nowhere else. Its tropical moist forest area in 1980 was estimated at 10.3 million ha on the basis of 1950s survey data, but a recent Landsat survey indicated that only 3.8 million ha remained in 1985 (Green and Sussman, 1990). With the third largest deforestation rate in Africa and an annual loss of 1.6 per cent of its forest Madagascar is a priority area for global biodiversity conservation (see Chapter 7). Elsewhere in East Africa little tropical moist forest remains, except for fragments in Uganda, Kenya and Tanzania.

Asia-Pacific

The Asia-Pacific region had 264 million ha of tropical moist forest in 1980, a quarter of the world total. Three quarters of this was in Burma, Indonesia, Malaysia and Papua New Guinea. Indonesia was the leader with 113.6 million ha, more tropical moist forest than any

Table 5.8. *Percentage annual deforestation rates for countries in the humid tropics 1976–80*

Ivory Coast	7.0
Nigeria	4.8
Thailand	4.0
El Salvador	4.0
Costa Rica	3.7
Honduras	2.9
Nicaragua	2.3
Guinea-Bissau	2.3
Brunei	2.2
Ecuador	2.1
Liberia	2.1
Colombia	1.7
Madagascar	1.6
Laos	1.6
Uganda	1.3
Equatorial Guinea	1.2
Malaysia	1.1
Philippines	1.1
Ghana	0.9
Sierra Leone	0.8
Vietnam	0.8
Panama	0.7
Belize	0.7
French Guiana	0.7
Indonesia	0.5
Cameroon	0.5
Dominican Republic	0.5
Brazil	0.4
Venezuela	0.4
Burma	0.3
Peru	0.2
Kampuchea	0.2
Zaire	0.2
Cuba	0.2
Bolivia	0.2
Central African Rep.	0.1
Gabon	0.1
Papua New Guinea	0.1
Guatemala	0.0
Guyana	0.0
Surinam	0.0

Source: Grainger (1986), based on Grainger (1983)

other country except Brazil. It also had the region's highest deforestation rate, 550,000 ha per annum in the late 1970s – just below the regional average rate of 0.6 per cent. The deforestation rate in Malaysia was less than half Indonesia's but of more concern since Malaysia has less forest. Deforestation rates in Papua New Guinea and Burma were apparently low but Burma's rate was probably underestimated by FAO/UNEP: according to a Landsat survey it was at least 600,000 ha per annum in the early 1980s (Allen, 1984); Myers (1989) put it even higher at 800,000 ha per annum later in the decade.

Of the five other countries, India, Kampuchea, Laos, the Philippines and Thailand, which accounted for 19 per cent of all tropical moist forest in the region, the last two countries are of most concern. The Philippines was once the region's leading log exporter but its forests are now depleted. There were only at most 9.3 million ha of forest left in 1980, according to the FAO/UNEP data used by Grainger (1983), giving a national forest cover of 31 per cent. But by the end of the 1980s national forest cover was down to 24 per cent compared with 60 per cent in 1939 (Figure 4.4). Remote sensing surveys suggest that FAO/UNEP's estimated annual deforestation rate of 100,000 ha for the late 1970s was too low and that the actual rate was nearer 250,000 ha, though this later declined to a mean of 121,000 ha for the 1980s (Kummer, 1992). Thailand had a similar forest area to the Philippines in 1980 but a higher deforestation rate (325,000 ha per annum in the late 1970s). Its annual percentage forest loss (4.0 per cent) was the highest in the world after Ivory Coast and Nigeria. National forest cover is now down to 16 per cent, compared with 53 per cent in 1961. It has become a net importer of forest products, like Nigeria.

Latin America

Latin America had 613 million ha of tropical moist forest in 1980, 57 per cent of the world total. Half was in just one country, Brazil, and another 31 per cent in four other countries, Bolivia, Colombia, Peru and Venezuela. Only 6 per cent was in Central America and the Caribbean Islands. Over half of all annual deforestation of tropical moist forest takes place in Latin America.

Of the 3.3 million ha per annum deforested in the late 1970s, 1.4 million ha was lost in Brazil, 800,000 ha in Colombia and 300,000 ha in Ecuador (Table 5.7). Ecuador and Colombia have the highest population densities in South America and both have lost a lot of the tropical rain forest that used to cover their Pacific coastal strips.

Myers (1989) suggested that Colombia's deforestation rate had declined in the late 1980s, with Ecuador holding steady and the rates in Bolivia and Peru more than doubling. He also gave a higher deforestation rate for Central America, which was worrying because even in the late 1970s El Salvador, Costa Rica, Honduras and Nicaragua were each losing more than 2.0 per cent of their forest every year.

There has been considerable debate about the rate of forest loss in Brazil, and for good reason since in the 1970s it accounted for over one fifth of the deforestation rate for the entire humid tropics and just a fractional increase in its rate can have a major impact on the global rate. Compared with the other two leading forest countries, Indonesia and Zaire, its forest area has been monitored more attentively over the last twenty years by its own remote sensing experts, but this has still not prevented some exaggerated figures emerging, and the impact of this on global deforestation rates has been discussed earlier.

Brazil's National Space Agency (INPE) estimated the annual deforestation rate in Amazonia between 1975 and 1978 as 1.565 million ha, based on a comprehensive comparison of Landsat images in those two years (Tardin et al, 1980). In 1989 half of Amazonia was studied with high resolution Thematic Mapper (TM) images collected in 1988. Compared with the 1978 survey the deforested area had increased by 17 million ha to 25.1 million ha, giving an annual deforestation rate of 1.7 million ha per annum for 1978-88 (da Cunha, 1989). This was later said to be an underestimate due to its partial coverage, neglect of small clearances and the obscuring effect of clouds (Fearnside et al, 1990). In 1990 a fully comprehensive survey with TM images from 1989 put the cleared area at 30 million ha in 1989, the deforestation in that year had been 2.6 million ha, and the average deforestation rate for 1978-89 was 2.1 million ha per annum with an error of ±10 per cent (Fearnside et al, 1990). This is now regarded as the best estimate of the deforestation rate for Brazilian Amazonia in the 1980s. According to the same survey 8 per cent of all tropical moist forest in the region has now been cleared, including the deforestation that took place before 1960.

An earlier estimate of 8.1 million ha for the deforestation rate in 1987 (Setzer, 1988) was later disowned by INPE (Marcio Barboso, INPE, personal communication). It was made using low resolution AVHRR satellite imagery to measure not changes in forest area but the number of fires in four states in the region. The results were then used to estimate a deforestation rate for all Amazonia, but without checking whether the fires (which are a regular pasture management feature) were in previously forested areas or not, and neglecting that a fire could overload the satellite sensor and so appear far larger than it actually was (da Cunha, 1989).

The critical countries

Deforestation is a problem for most countries in the humid tropics but a major problem for 11 in particular: Ghana, Ivory Coast, Madagascar, Nigeria, Uganda, the Philippines, Thailand, Vietnam, Cuba, the Dominican Republic and El Salvador. Some of these countries were referred to above, but in all of them deforestation had reached, or was approaching, critical levels by 1980 according to one or all of the following indicators: percentage national forest cover, percentage annual deforestation rate and forest area per capita (Table 5.9).

Table 5.9. *Countries at critical levels of deforestation in 1980*

	Percentage national forest cover	Percentage annual deforestation rate	Forest area per capita (ha)
Ghana	8	1.1	0.15
Ivory Coast	14	7.0	0.54
Madagascar	18	1.6	1.18
Nigeria	7	4.8	0.07
Uganda	4	1.3	0.06
Philippines	26[1]	1.1[1]	0.17
Thailand	18	4.0	0.18
Vietnam	27	0.8	0.14
Cuba	13	0.2	0.13
Dominican Republic	16	0.5	0.08
El Salvador	7	4.0	0.02

1: Corrected from that cited in the source which is now known to have overestimated the forest area and underestimated the deforestation rate.
Source: Grainger (1983, 1986)

Improving the monitoring of tropical deforestation

This chapter has shown that global and national estimates of deforestation rates are both very inaccurate, and that while they give some idea of the overall scale of the problem and of differences between rates in various countries it is difficult to track long-term trends at global or national level, except for a few countries and even then great care is needed. This will probably make all readers very sceptical when any 'new' estimate of deforestation rates is announced in future, and perhaps that is not a bad thing, but why have we got into this sorry state of affairs and how can we get out of it?

The two simplest reasons for our problem are that monitoring global environmental change has not been taken seriously until recently, and that during this century most tropical countries have only surveyed their forests at infrequent intervals. All estimates of the global deforestation rate discussed in this chapter came from aggregating national estimates, rather than being measured by a global monitoring programme. That we have any global estimates at all is due to the hard work of FAO and independent experts like Norman Myers in collecting data from many countries. FAO has compiled international statistics on forest areas and land use since it was founded at the end of the Second World War. It has traditionally relied on data supplied by governments rather than collecting its own data by direct monitoring, but it halted preparation of the 1972 World Forest Inventory (previous inventories had been published in 1950, 1955 and 1965) because it was dissatisfied with the quality of data received.

The first recognition of the need for direct global monitoring came at about the same time that FAO had reached this watershed in its world forest inventory work, when the UN Conference on the Human Environment, held in Stockholm in 1972, agreed in Resolution No. 25 on the need for a continuous monitoring system for all world forest resources, with a prominent satellite component. This was impressive given that the first Landsat satellite had only been launched in that year. The new UN Environment Programme and FAO decided to collaborate on implementing the resolution and commissioned a feasibility study in 1973. Two years later this recommended launching a project to make a baseline survey of tropical forest cover (using satellite imagery as a main source of data), assess changes in tropical forest cover over the past ten years and prepare guidelines for a continuous monitoring system (FAO, 1975).

Though the project was implemented the results were rather different from what had been anticipated in the feasibility study. For instead of actively monitoring tropical forests FAO followed its usual practice of collecting as much existing data as it could from governments and other sources. It interpreted these data with discretion but only commissioned new surveys, mostly using Landsat imagery as the main data source, when other data (which included a comprehensive side-looking airborne radar survey of Brazilian Amazonia and other remote sensing surveys) were insufficient. The massive report of the project, summarized above, was a great step forward. But although its 1500 pages were packed with figures they lacked the accurate estimates of deforestation rates that were so badly needed. Estimates of forest areas were also unreliable and did not provide the baseline satellite-based survey that would be an essential foundation for a future continuous monitoring programme (Grainger, 1984). Indeed, the whole concept of continuous monitoring was denied when the project team was disbanded after the report's publication. When FAO decided to undertake another assessment for 1990 it had to assemble a new team.

Global monitoring and local monitoring

The notion of global monitoring has proved surprisingly controversial. Even people who acknowledge the importance of the tropical deforestation issue still seem quite content with fuzzy, inaccurate figures. Some would argue that we do not need better global data and that a more 'bottom-up' project-based approach is the way to solve the problem. Others believe that individual countries should monitor their own resources, perhaps also claiming that global monitoring by a UN agency infringes national sovereignty over data on land resources, and that effective satellite monitoring is impossible anyway without good local ground data collection. While the latter is undeniable, global monitoring and ground data collection are certainly not incompatible.

But none of this fully explains why we have not monitored tropical deforestation properly in the past and are still hesitant about doing so in the future. The establishment of a continuous global satellite monitoring system has been prevented so far by a combination of institutional, political and financial constraints, but could it be that the main reason for our inaction is that we simply do not think our planet worth monitoring in its entirety? How else can we explain the spending of billions of US dollars on space telescopes and satellites to monitor other planets and galaxies? Humanity must realize that if, as a species, we have now reached the stage of being able to modify the global environment, and are threatening to do so imminently through the greenhouse effect, we have an absolute duty to monitor our impacts, both to document them and to provide the basis for better management in future. The sovereignty issue should be clarified by a formal exploration, under United Nations auspices, of the issues involved, for it is certainly a sensitive matter for a number of governments, even though satellite imagery of any country is now freely available to anyone who can pay for it. Perhaps one answer is to separate data needed for global purposes, such as national forest areas and deforestation rates, from more sensitive higher resolution data with economic connotations, such as the commercial timber content of forests.

As most tropical countries still have only limited remote sensing capabilities, it is quite understandable if they give a lower priority to collecting data for global monitoring programmes than those needed for specific national purposes, such as: supporting national land suitability assessments, identifying areas of fertile land suited to intensive agriculture and areas of degraded land needing rehabilitation, and keeping an eye on areas critically at risk from deforestation. The list is endless, and most such imperatives would contribute to a national programme to improve environmental management and control deforestation, as suggested in Chapter 7. An effective global monitoring programme would not duplicate such activities, but could receive data from them at the discretion of governments, in addition to the data it collected itself. It would also benefit local action financially by raising international awareness of the need to control deforestation and improve forest management.

What should be the way forward for global monitoring? In the light of the previous paragraph it seems unlikely that a decentralized global monitoring programme that merely compiles data collected by a large number of national monitoring schemes, will produce estimates of the quantity, quality and frequency needed in the foreseeable future. National monitoring activities will take time to develop and even then most countries will probably not focus on producing data of the right kind and frequency required for global monitoring. On the other hand, a centralized continuous global monitoring programme, in which a single agency interprets satellite imagery, other remote sensing data and ground truth data to monitor the state of the tropical forests worldwide on a continuous basis, with links to collaborating scientists in individual countries, appears to be the only feasible way to ensure that we can obtain yearly estimates of forest areas, deforestation rates and environmental impacts (such as degradation of biomass and biodiversity) before the year 2000 or even 2010.

It would be best if the UN could retain responsibility for global forest monitoring, whether the agency involved were FAO, UNEP, or perhaps a new one dedicated to this task. But there is no guarantee at the moment that the team producing the FAO 1990 Tropical Forest Resource Assessment will stay together as the basis of a continuous monitoring programme for the tropical forests. If so, and no other UN agency takes on this role, it is possible that a non-governmental organization or research institute could take on this important job (Grainger, 1990e). There are now few technical impediments to global monitoring, given the availability of satellite imagery, computing equipment and software. What has prevented effective global monitoring so far has been a lack of will and institutional obstacles relating to the ability to organize monitoring on such a huge scale. If these can be overcome, by the UN or other organizations, the financial obstacles should be minimal. Global monitoring is an idea whose time has come. It is an obligation of a human race that has started to modify the global environment and is now seeking ways to curb its worst excesses and manage the planet in a better way.

The monitoring imperative

However much we appreciate the seriousness of what is happening to the tropical rain forests we lack the vision and the will to make full use of existing remote sensing technology to monitor the changes we are making. Consequently, great uncertainty attaches to estimates of current deforestation rates and even more so to historical trends. Much less is known about rates of forest degradation and trends in carbon storage in biomass and net carbon emissions. The annual deforestation rate in the tropical moist forests has been variously estimated at 11–15 million ha in the early 1970s, 6.1 million ha in the late 1970s, and 12.2–14.2 million ha in the 1980s. But given the errors involved in making those estimates it is always possible that the rate did not vary all that much, perhaps staying close to 10 million ha per annum over the whole period.

Looking into the future, if 1,081 million ha of tropical moist forest remained in 1980 and the global rate of deforestation continued unchanged at 6.1 million ha per annum all the forest would be lost by the year 2157, but if instead the rate were 14.2 million ha per annum and only 800 million ha remained in 1990 the critical year would be 2047. The comments in Chapter 4 about the limitations of linear projections should be borne in mind, but even so we can see the effect that inaccurate estimates of deforestation rates and forest areas can have on future planning.

Given the importance of tropical deforestation in its own right and as a key element of global environmental change it is tragic that we are in such a state of ignorance. Tropical deforestation is so important that it deserves to be continuously monitored by satellites and other means on a year-to-year basis. Until this happens future estimates will remain very inaccurate, programmes intended either to control deforestation or to understand its environmental effects will be severely flawed, and the pressure to take action to control deforestation will be much weaker than it should be. The current state of uncertainty should give as much cause for concern as the loss of forests itself.

6

The environmental effects of deforestation and logging

There is no more powerful or straightforward reason for wanting to conserve the tropical rain forests than that their removal would destroy one of the planet's most valuable ecosystems and a priceless part of our natural heritage. But we can go further than this, and point to the threat that deforestation and logging pose to the rich biological diversity of tropical rain forests, which are thought to contain half of all the world's species of plants and animals. Problems also arise because of the soil degradation, changes in water flows and increased sedimentation of rivers, reservoirs and irrigation systems caused by deforestation and logging. Finally, owing to the carbon dioxide released into the atmosphere after the burning of cleared vegetation, deforestation contributes to the greenhouse effect that is expected to lead to climate change on a global scale that could be felt by everybody in the world within decades. This chapter looks at these three major environmental effects of deforestation and logging.

The threat to biological diversity

Our planet contains at least 10 million species of plants and animals and probably a great many more – as many as 30 million if some projections of the number of insect and other invertebrate species are correct, though there is a lot of uncertainty about this (Erwin, 1983; Wolf, 1987). Each day most of us interact with just a few of these species in the food we eat, our domesticated animals, the plants in our gardens, and wild birds and insects.

It is hard to imagine what millions of species mean in practice, but they are just one aspect of the immense biological diversity of our planet, commonly shortened to 'biodiversity'. This has three main components: ecosystem diversity, species diversity, and genetic

146

diversity. So the biodiversity of a given area will be characterized by the presence of different kinds of ecosystems; its species diversity, including the number of species and the number and distribution of endemic species – those limited to a particular country, province or locality; and its contribution to the genetic diversity of particular species, which is very important to plant breeders, and which allows a wide range of characteristics among individuals, sometimes including distinct sub-populations, each with its own geographic range and genetic traits.

The tropical rain forests contain about half of all species in the world and necessarily the number of species per unit area is very high. On the global scale species density increases markedly from the Poles to the Equator, where it reaches a pinnacle in the tropical rain forests. Compare the island of Britain, for example, which has 35 native tree species, with the small state of Brunei on the northwest tip of the island of Borneo in Southeast Asia, which has two to three thousand tree species but is only one fortieth the size of Britain.

Deforestation, therefore, puts at risk not only the biodiversity of the humid tropics, but also that of the world as a whole. It is already increasing the rate of species extinction so that plants we are using, or could use in the future, will disappear, as will many animals too. Some plants will vanish even before they have been catalogued, others will go when their potential uses are known but before they have been domesticated. At the same time the genetic diversity of many individual species is also being degraded. As some of these species are commercially valuable crops, such degradation will threaten our ability to continue breeding new varieties to keep ahead of the pests and diseases that threaten production.

Biodiversity in the humid tropics

Tropical rain forest is one of the world's 12 major types of ecosystem, or biome types (Whittaker, 1975). Others include temperate forest, boreal forest and Mediterranean vegetation (Figure 6.1). It is not a homogeneous biome type and can be divided first of all into three distinct regional blocks, or biomes, in Africa, Asia-Pacific and Latin America. Each biome has its own characteristics and species composition and comprises a patchwork of different sub-types of tropical rain forests, of which fourteen were listed in Chapter 1.

The tropical rain forests are thought to contain between two and five million species of plants and animals: half of all species on the planet and two thirds of all tropical species (Myers, 1985b). As we have catalogued only a small proportion of these species the final

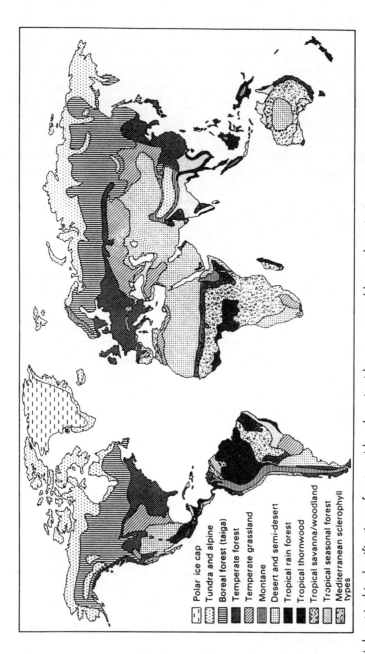

NB Semi-desert in this classification refers to arid and semi-arid areas covered by scrub vegitation

Sources: Based on the map and classification of Whittaker (1975), with modifications using data from Eyre (1968), Goudie (1984), Grainger (1986a) and Walter (1973)

Figure 6.1 *The major biome types*

Legend:
- Polar ice cap
- Tundra and alpine
- Boreal forest (taiga)
- Temperate forest
- Temperate grassland
- Montane
- Desert and semi-desert
- Tropical rain forest
- Tropical thornwood
- Tropical savanna/woodland
- Tropical seasonal forest
- Mediterranean sclerophyll types

total could be much higher, and this would raise the world species total too. Only 1.4 million species of plants and animals have yet been recorded and of these about 500,000 are tropical (Raven, 1987a). Over 150,000 of the 240,000 species of flowering plants grow in the tropics (Raven, 1976), 60 per cent of these in Latin America (Gentry, 1982).

Threats to biodiversity

Deforestation, logging and other human impacts threaten biodiversity in three main ways. First, they generally reduce the number of species, the genetic diversity of individual species and the variety of ecosystems. Second, they degrade the genetic diversity of key domesticated crop plants. Third, they threaten the survival of species already exploited in their wild form in the forests – the so-called minor forest products. We now discuss each of these in turn.

The oldest threat to species is overhunting. Many species are now at risk due to this cause: only a few hundred Sumatran rhinos, for example, are left in Sumatra, Burma and Peninsular Malaysia. Some species of rhinos, tigers and birds have already been hunted to extinction. Deforestation for permanent agriculture destroys habitat and replaces it with a more extreme environment that many species cannot tolerate. An estimated 60 per cent of birds breeding in Peninsular Malaysia, for example, cannot survive outside the forest (Wells, 1971, 1974), and the number of songbirds from the USA who migrate to the forests of Central America every winter has dropped as the forests have been cleared (Keast and Morton, 1980).

The impacts of shifting cultivation and selective logging fall between these two extremes. Traditional shifting cultivation, with its small scattered clearances and long rotations, should have little long-term effect, as primary forest trees will easily invade the abandoned plots. But the impact becomes greater as rotations get shorter and only bushy forest fallow can regenerate before the next clearance. Selective logging, on the other hand, risks depleting commercial timber species, introducing other species characteristic of secondary forests and, by modifying or removing animal habitats and food sources, having detrimental effects on animal populations. There are major consequences for plants that depend on specific animals for pollination or seed dispersal. *Virola surinamensis*, an important commercial timber tree in South America, depends for seed dispersal on just six species of birds and one monkey. If one or more of these should become extinct the tree's future would be in jeopardy.

149

All forms of disturbance, however temporary, displace animals from part or all of their territories. The more extensive and prolonged the disturbance, the greater the chance that animals in an area will become crowded and their populations decline owing to social pressures, limitations on food supplies and impaired reproductive ability. By removing just a few plants from the forests selective logging could disrupt the complex annual calendars of food sources that enable many rain forest animals, like the orangutan, to cope with the low density and irregular flowering and fruiting regimes of the plants they eat. Other plants may be affected if animals on which they rely for pollination or dispersal leave to search for food elsewhere.

The size of the threat

The overall effect of deforestation, logging and overhunting on biodiversity is still unclear. Assessments initially concentrated on threats to particular species but were later extended to estimate current and possible future extinction rates. The International Union for the Conservation of Nature (IUCN) has produced lists of threatened species, grading them as Intermediate, Rare, Vulnerable and Endangered in increasing order of threat. Endangered species, which risk becoming extinct if present conditions continue, include the koupray or wild cow from Southeast Asia, of which only 100 individuals remain, the broad-nosed lemur from Madagascar, which is down to just two colonies, and the southern bearded saki, a monkey living in the forests of northeast Amazonia.

Extinction is an irreversible change and once a species has gone it it is lost for good. No good wishes, no high technology can bring it back. Extinction occurs naturally at a rate of about one species every two years, balanced by the evolution of successor species better adapted to existing conditions (Wilson, 1989). But the human race has a great propensity to increase the extinction rate, and nowhere is more vulnerable in this respect than tropical rain forest.

While a species actually becomes extinct when its last member dies (Goodman, 1987) it is put at risk much earlier when its numbers drop so low that it could be eliminated by drought, disease or other random events. Ecologists have been trying to understand more about how low the populations of individual species can fall before they become susceptible in this way. Knowing the size of this 'minimum viable population' is vitally important for effective conservation planning (see Chapter 7) (Soulé, 1980; Gilpin and Soulé, 1986).

The environmental effects of deforestation and logging

Quantifying the impact of tropical deforestation on biodiversity is difficult, for apart from rhinos and some other large animals there are few documented examples of extinctions in continental parts of the humid tropics. Most evidence is from islands such as Hawaii where a long history of isolation and separate development led to a largely endemic flora and fauna. Hawaii lost 41 native bird species before European colonization in 1778, and since then 23 bird species and 177 plant species have vanished due to deforestation and the effect of exotic plants and animals (Olson and James, 1984; Nature Conservancy, 1987). A further 30 bird species and 680 plant species are endangered today.

Available estimates of current extinction rates in the humid tropics are still little more than guesses, but increasingly rigorous methods are being employed to tackle this question. For a long time in the 1980s the 'accepted' figure was at least one species lost every day (Myers, 1979a; Raven, 1987b). More recently, Edward O. Wilson, of Harvard University, raised this to 11–16 species a day, though this assumed an annual reduction in tropical rain forest area of 1 per cent (see Chapter 5) (Wilson, 1989). Even with a more conservative 0.5 per cent annual forest loss rate, of the same order as that estimated for the late 1970s (Grainger, 1983), five to eight species could become extinct every day.

More sophisticated estimates are made possible by applying the theory of island biogeography, which states among other things that the equilibrium number of species on an island is proportional to its area. So as the size of the island falls the number of species should decline too (MacArthur and Wilson, 1967). By treating areas of tropical rain forests as huge 'islands' surrounded by 'oceans' of cleared land, Daniel Simberloff predicted the loss of 15 per cent of all plant species and 12 per cent of bird species if the area of tropical rain forest in Latin America were to fall by 60 per cent between 1980 and 2020 (Simberloff, 1986). Walter Reid and Kenton Miller, of the World Resources Institute, used this method to test the effect of a less drastic cut in the area of all tropical closed forests, assuming in their Low Scenario that deforestation rates between 1990 and 2020 continued unchanged at the levels projected for 1981–5 by FAO/UNEP (see Chapter 5) but were twice as high as this in the High Scenario. The result was a loss of 6–14 per cent of all species in Africa, 7–17 per cent of all species in Asia and 4–9 per cent of all species in Latin America, depending on the scenario (Reid and Miller, 1989). Better estimates of present and potential extinction rates depend on gaining more accurate data on species numbers and deforestation rates, and improving our understanding of how deforestation and logging affect biodiversity.

Present and future crops at risk

Too great a focus on species extinction can divert our attention from other effects of deforestation that degrade the genetic diversity of individual species rather than removing them totally. This is a major threat to the genetic resources of several key economic crops that originated in the humid tropics and still grow wild there today. These include brazil nut, cashew nut, cocoa, groundnut, passion fruit, pineapple, rubber and tapioca (from Brazil); papaya and sweet potato (from Peru); oil palm and sesame (from West Africa); and bamboo, banana, cardamom, ginger, rice and yam (from Southeast Asia) (Grigg, 1977).

Crops now grown in fields or plantations outside tropical rain forests have been selectively bred from wild plants to give the best yields under particular environmental conditions and so make use of only some of the genetic characteristics from the entire population, or 'gene bank'. It is vital to retain a wide genetic diversity of wild plants so that plant breeders can counter threats to crop productivity caused by new pests and diseases and changing climate. Since the entire human population depends – as producers, processors or consumers – on a very small number of crops the economic consequences of being unable to continue cultivating them could be horrendous. Breeders have to resort to the gene bank of the cacao plant (the source of cocoa) in Amazonia in the fight against swollen shoot disease and huge sums of money are spent in breeding new groundnut varieties resistant to diseases such as leafspot (Caufield, 1984). Some high yielding rice varieties last only two years before being attacked by a new insect pest and needing replacement (Madeley, 1990a).

While maintaining the wild populations of crop plants in tropical rain forests is important another problem is that as single high yielding varieties become adopted over increasingly larger areas not only do they become more susceptible to attack by pests and diseases, but also the previous diversity of cultivated varieties is drastically cut. These also represent an important part of the genetic diversity of the crop. When in the 1970s plant breeders anxious to breed a new high yielding rice variety were searching for a traditional strain that could withstand high winds they eventually found one in Taiwan, but it had been almost totally replaced there by one of the early high yielding varieties (Myers, 1978).

The environmental effects of deforestation and logging

Deforestation also threatens the genetic diversity of plants that are not yet major crops but could become so in the future. They include uvilla (*Pourouma cecropiifolia*), a medium-sized tree from western Amazonia whose tasty fruits can be used to make wine (Prance, 1982a; Balick, 1985); the serendipity berry (*Dioscreophyllum cumminsii*) from West Africa, which is 3000 times sweeter than sucrose but whose active ingredient is a protein instead of a high calorie sugar (Myers, 1985a); and a primitive relation of maize (*Zea diploperennis*), found in a montane forest in west-central Mexico, which is a perennial rather than annual plant, grows in very wet conditions and is resistant to a number of viruses that attack maize (Myers, 1981).

The threat to minor forest products

The third threat of deforestation and logging is to the wide range of non-timber products, also known as minor forest products, that are harvested in the forest. Local people gather nuts and other fruits; hunt wild game, which accounts for at least a fifth of animal protein intake in many West and Central African countries; and also collect a variety of gums, resins, oils, tannins, latexes, fibres, bamboos and medicines. Some are used for subsistence purposes, but others are sold on local and world markets and are a major source of income. Indonesia's minor forest product exports in 1985, for example, were worth $154 million, or 12 per cent of all its forest products exports (Gillis, 1988). In other countries the proportion is far higher.

Gums, now largely replaced in their former use as adhesives by synthetic substitutes, are still very important in food and cosmetic manufacture. For example, chicle, the main ingredient of chewing gum, is tapped from a tree in the tropical rain forests of Guatemala. Resins from trees of the genera *Agathis* in Southeast Asia and Copaifera in Amazonia are used to make varnishes and earned Brazil alone $500,000 in export revenue in 1985 (Prance, 1990). Most edible oils, such as palm oil and coconut oil, are now grown on plantations (see Chapter 2) but 'essential oils', such as sandalwood, citronella, camphor and cinnamon, are still extracted from tree leaves, barks and other plant parts to be used in the manufacture of perfumes and food flavourings, with a world trade volume of over $1000 million in 1980 (Duke, 1981; Princen, 1979; Pryde et al, 1981). Tannins have been used for centuries to cure leather and though synthetic substitutes are now common, natural tannins from the barks of trees in mangrove forests are still used in this way and also to make industrial lubricants.

Controlling tropical deforestation

Rubber, made from latex tapped from the rubber tree *Hevea brasiliensis*, is a prime example of a minor forest product that became a major world crop. Wild rubber trees are still tapped in Amazonia but the world market is now supplied mainly from highly productive plantations in Southeast Asia. *Hevea brasiliensis* is just one of many latex-producing trees but became the most popular for giving a fresh supply quicker than any other. Conservation of wild trees in Amazonia is important to support the future breeding of new varieties, including some that are resistant to the pests that caused the failure of previous Brazilian plantations.

Tropical rain forest trees are rich sources of fibres with a range of uses. Brushes are made from the stems of palm trees and mats from coconut husks. Rattans, the climbing vines found in Asian tropical rain forests, are used locally for many purposes and are the basis of a large international furniture industry. In 1985 Indonesia exported cured stems of rattan palms (*Calamus* spp.) worth $97 million, but the following year stem exports were banned by the government so as to focus on the export of furniture and other higher value products instead (Priasukmana, 1988). The hundreds of species of bamboo grasses found in tropical rain forests are utilized in the manufacture of furniture, baskets, paper and other products, and on building sites in Asia there is a good chance that bamboo scaffolding will be used.

The last group of minor forest products, and deserving of special attention, are the forest medicines. Chemicals extracted from tropical rain forest plants may literally both kill and cure. Curare is obtained from the bark of the Brazilian climbing vine *Chondodendron tomentosum* and was traditionally used by Amazonian Indians to give a poisonous coating to their blowpipe darts. But the main constituent of curare, tubocurarine, was found in 1942 to be a powerful surgical anaesthetic. Ouabain is another dart poison obtained from an African climbing plant but is now also used as an emergency heart stimulant (Humphreys, 1982).

Some chemicals have long been used as medicines by forest dwellers. The alkaloid quinine was one of the first treatments and preventatives for malaria. Now mainly grown on plantations it originally came from the bark of the tree *Cinchona officinalis* in the forests of Ecuador, Peru and Bolivia (Salati et al, 1987). Reserpine, extracted from Indian snakeroot, *Rauvolfia serpentina*, is another alkaloid and was a traditional medicine for thousands of years before becoming the first modern tranquillizer. While there are now synthetic substitutes, extracts from *Rauvolfia* are still the main source of reserpine and retail sales of derived tranquilizers are worth hundreds of millions of dollars in the USA alone (Myers, 1985a).

The environmental effects of deforestation and logging

The commercial value of steroids used in contraceptive pills is even higher. Diosgenin, which initially came largely from Mexican wild yams, *Dioscorea composita* and *D. floribunda*, is used to produce a sex hormone that overcomes female sterility and to manufacture other steroids, including those in contraceptive pills and cortisone and hydrocortisone, which are used to treat various diseases. Another source of diosgenin is the West African calabar bean.

There are hopes that among the huge number of chemicals in tropical rain forest plants are some that could cure cancer. Already two alkaloids obtained from the rosy periwinkle, *Catharanthus roseus*, originally native to the Caribbean but now cultivated more widely, have achieved worldwide prominence for their role in the treatment of leukaemia. Thousands of other plants are being screened for alkaloids that might make them valuable anti-cancer drugs (Humphreys, 1982; Booth, 1987).

There is a great danger that in many countries the future of minor forest products from tropical rain forests is being threatened by deforestation and overlogging. With good management they could be supplied as sustainably as timber and make a major contribution to the national economy and to export revenue. So while we should undoubtedly protect tropical rain forest for its own sake, the benefits accruing from some minor forest products, like the extracts from the rosy periwinkle, show that conservation could be in our own interests too.

The effects of deforestation and logging on soil

Clearing or logging forest exposes soil to erosion and to four other forms of degradation: fertility depletion, compaction, laterization and landslides. The poor fertility of most soils in the humid tropics is further reduced when overcultivation depletes soil nutrients and the organic matter crucial to maintaining a good soil structure and protecting it from erosion. The effects of soil degradation become apparent locally in the lower productivity and sustainability of farming on degraded soils, and more widely in the sedimentation of rivers, mentioned below.

Soil erosion and land use

Rainfall in the humid tropics is not only high and continuous throughout the year but also very erosive, arriving in brief, heavy showers (Douglas and Spencer, 1985). Fortunately, rain is slowed down by the dense vegetation cover of tropical rain forest, while the surface cover of litter and herbaceous vegetation protects soil from the impact of raindrops and, together with the dense network of shallow tree roots, from being washed away by surface water. Some soil erosion happens all the time within the forest but the erosion rate increases sharply if forest cover is cleared or disturbed. The loss of fertile topsoil increases the sediment load of streams and rivers and reduces the ability of land to grow crops and support other vegetation.

The size of the erosion hazard varies with the land use replacing forest. Shifting cultivation should have a low impact, as only small areas are cleared and cropped at any one time, clearing is by hand rather than with mechanical tools, many tree roots remain under the soil surface to give stability, and any soil washed away cannot travel too far. One experiment in the Philippines found that the annual rate of soil erosion increased from 0.73 tonnes per ha under primary forest to 5.3 tonnes per ha when forest was cleared by shifting cultivators to grow rice. On shifting cultivation land colonized by *Imperata cylindrica* grassland (see Chapter 2) the loss rate was 1.4 tonnes per ha (Kellman, 1969).

Erosion is much worse when forest is replaced by permanent field crops. An annual loss of 70 tonnes per ha was found on cultivated land in the Ivory Coast with only a 7 per cent slope (UNESCO/ UNEP/FAO, 1978). The greater the slope, the greater the erosion rate, and over 1000 tonnes per ha per annum was lost on land with a moderate to steep slope in Sabah used for annual field crop cultivation (Brunig et al, 1975). Mountainous areas are less common in Africa and Latin America, but even on the extensive plateaux in these regions there is a considerable risk of soil erosion since the land is more like a rolling plain than perfectly flat.

The amount of erosion also depends on how the land is cleared. Removing forest in Nigeria with crawler tractors and then ploughing it caused four times as much erosion as when it was manually cleared and ploughed, and 50 times more than on land manually cleared and then cropped without ploughing (Lal, 1981). The widespread use of mechanical clearing in the Indonesian Transmigration Programme, for example, cut costs and allowed contractors to keep to schedule, but it also removed a lot of soil fertility even before the first crops had been planted (Goodland, 1981).

Much less soil erosion occurs under tree crop plantations. But a survey of research in this field found that while the erosion rate only doubles when forest is replaced by a tree crop plantation that retains some form of surface vegetation, such as a cover crop or mulch, if the land beneath the trees is kept clean from weeds 15 times as much soil is lost. A similar pattern is found in timber plantations, but as they mature and undergrowth is no longer weeded the erosion rate should approach that found under natural forest (Wiersum, 1984).

Soil compaction

Another type of soil degradation is compaction, in which a hard surface crust forms when soil comes under heavy pressure from machinery or animals, or crumbs of soil are converted by mechanical cultivation into a thin powder, which is packed together by the impact of raindrops. The heavier the load the deeper the compaction. This is yet another risk associated with mechanical land clearing methods, but it also affects livestock raising schemes, for though experimental pastures have low soil erosion rates, once they are grazed the constant trampling by animals' hooves compacts the soil. Surface water run off is greater, and this in turn increases the loss of soil by water erosion (Suarez de Castro and Rodriguez, 1955).

Logging and soil degradation

The main erosion risk from selective logging occurs during and shortly after logging when the network of skid trails and logging roads act as convenient channels for surface water runoff and soil erosion. An annual erosion rate equivalent to almost 155 tonnes per ha was measured on a new skid trail in Indonesia, but this had halved two years later when vegetation had regrown on the trail (Ruslan and Manan, 1980). The soil on skid trails is also compacted by crawler tractors. Governments aim to minimize erosion by setting strict standards for the construction of logging roads, but if concessionaires try to save money and fail to comply with these standards then substantial erosion can occur. If proper drainage is not installed the build-up of water along the sides of roads could eventually wash away huge chunks of road.

Laterization

It used to be assumed that deforestation and subsequent overuse of soils in the humid tropics would lead to widespread laterization, turning large areas into barren deserts covered by what look like dark red concrete pavements (McNeil, 1964). But only some oxisols, ultisols and alfisols, covering just 2 per cent of the humid tropics in areas with a pronounced dry season, are now thought to be vulnerable in this way (Moorman and Van Wambeke, 1978). Laterization is a progressive phenomenon rather than a once-and-for all change. It begins with a layer of soft mottled material called plinthite, formed by deposition of iron and aluminium oxides when water is drawn up from deeper soil levels and evaporates from exposed surfaces in the dry season. Further drying leads to the formation of laterite, first as individual nodules, and later developing into a continuous layer of sub-soil blocks that impede drainage and root growth. Only if the sub-soil is exposed by topsoil erosion will these harden into a brick-like surface crust. Red laterite pavements are found in parts of Africa and monsoonal Asia but laterite should not form if the soil stays moist, as it does over a large part of the humid tropics.

Landslides

Landslides occur frequently in the humid tropics for entirely natural causes but one type, the mudflow, is particularly exacerbated by deforestation. Mudflows happen when soils with a high clay component are turned into a liquid slurry during intense rainstorms and then flow downhill. A serious mudflow occurred in southern Thailand in November 1988 after a very heavy storm in which 1,022 mm of rain fell in four days on a catchment with a 25 degree slope. The catchment was previously forested but had been converted into a rubber plantation. Because the land underneath the trees had been weeded this left only a sparse surface vegetation and the poor soil protection in the harsh conditions led to the mudflow (Rao, 1988).

Minimizing soil degradation

Some soil degradation is inevitable whatever land use replaces forest, but it can be minimized by employing good management techniques and matching land uses to land suitability, e.g. paying particular attention to soil fertility and the slope of the land. Recommended techniques include covering fields by some form of vegetation for a large part of the year, using manual rather than mechanical land clearing methods, leaving some surface vegetation in tree crop and timber plantations, and constructing logging roads and skid trails to good standards. Owing to the generally poor fertility of soils in the humid tropics soil degradation will always be a serious threat to the sustainability of agriculture and forestry. For this reason forest management, shifting cultivation, tree crop plantations and other agroforestry systems may well turn out to be the only sustainable land uses for large areas of the humid tropics.

Changes in water flows

Given the huge volume of water cycled annually through the humid tropics by rainfall, evaporation, transpiration and drainage by streams and rivers, the role of forest cover in regulating the flows of rivers and streams from upland catchments is extremely important. Removing the forest threatens these valuable environmental services and can have serious repercussions. But the popular notion of catchment forests acting like 'sponges' to soak up rainwater and then release it slowly is mistaken.

Forests help to maintain even water flows from catchments by returning most of the incident rainfall to the atmosphere and minimizing rapid runoff. About a fifth of all rainfall is intercepted by the canopy and then evaporates back into the atmosphere (Leopoldo et al, 1982). Some of the remainder runs down the trunks of trees but most falls from the canopy to the ground, its impact on the soil lessened by the canopy. Splash erosion is further reduced by a good covering of surface litter, which is actually of critical importance in minimizing soil erosion. Soil organic matter, tree roots and soil fauna keep soil porous so that as much water as possible infiltrates the soil instead of running off along the surface. A high proportion of infiltrating water is then taken up by tree roots and returned to the atmosphere by transpiration, leaving only a fraction of incident rainfall to run off underground. A study in Peninsular Malaysia found that 70 per cent of the annual rainfall of 2500 mm was lost by evapotranspiration and three quarters of this by transpiration (Kenworthy, 1970).

On well regulated catchments only a minor proportion of incident rainfall drains to the lowlands, and just a small fraction of this flows on the surface. A forested catchment should not release more than 5 per cent of total rainfall by surface runoff – the route known as overland flow (Figure 6.2). Most runoff is either through the soil (as quickflow) or more slowly through the bedrock (as groundwater flow). In the Malaysian study, of the 30 per cent of water not lost by evapotranspiration 26 per cent flowed away underground in the soil or bedrock and only 4 per cent ran off along the surface (Kenworthy, 1970).

By reducing the amount of water lost by evapotranspiration and infiltration, deforestation increases overall runoff, and a high proportion of this flows along the surface rather than underground. This, rather than the sponge mechanism, is how deforestation leads to faster and less evenly spaced runoff from catchments. For not only is there more water flowing downhill, but it is also released more quickly than from a forested catchment where most runoff travels slowly underground. The delay between a rainfall event and when most of the water reaches nearby rivers is drastically cut, and this

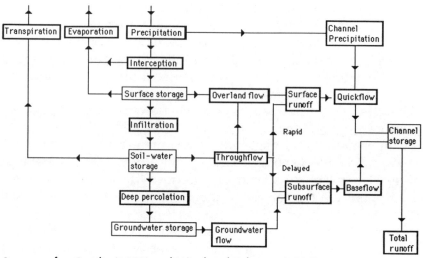

Sources: after Goudie (1989) and Ward and Robinson (1990)

Figure 6.2 *Water flow mechanisms in a drainage basin*

'bunching up' of draining water results in more uneven flows over the course of a year and a greater possibility of flooding.

The effects of different land uses

As with soil degradation, the impact of deforestation on water flows varies with the land use that replaces the forest. Only a limited number of empirical studies have been conducted so far in the tropics (Hamilton and King, 1983) but some generalizations can be made. A permanently cultivated catchment will have a much greater runoff than one still under forest, so when a forested catchment in Peninsular Malaysia was almost completely cleared for annual cropping, for example, overall runoff rose by 10 per cent (Low and Goh, 1972). Traditional shifting cultivation should have a minimal effect, but as rotations are reduced and a larger proportion of the catchment is under cultivation at any one time, with the remainder covered by low bushy forest fallow, the impact on water flows will approach that of permanent cultivation.

Tree crop plantations may be one of the more sustainable alternatives to forest but their water control functions are inferior: for example, the volume of water flowing through the soil, rather than the bedrock, from a catchment in Johore, Malaysia mainly covered by rubber and oil palm was double that from a forested catchment (Anon, 1971). Replacing forest by timber plantations will

160

give a higher overall runoff initially but this will decline as the plantations mature (Hamilton and King, 1983). Under experimental conditions the overall runoff from a catchment covered by pasture is often only slightly greater than that from a forested catchment (Queensland Department of Forestry, 1977) but on actual pastures, which are usually overgrazed with a degraded vegetation and compacted soil, runoff is likely to be far higher (Pereira, 1973).

The impact of selective logging on water flows depends upon how much of the forest canopy and other vegetation is removed and how logging is carried out. Heavy machinery will compact soil on skid trails and these, together with the logging roads, will act as channels for surface runoff. Water flows will therefore increase initially, declining later as vegetation regrows on skid trails. If gullies form on skid trails and poorly constructed logging roads then substantial soil erosion can occur, leading to high sediment loads for streams originating in logging areas, such as those reported from the Indonesian province of East Kalimantan (Hamzah, 1978).

Deforestation and flooding

It is often assumed these days that when a tropical river floods deforestation has played some part in it. But floods do occur naturally in high rainfall areas when rainfall and river flow rates (or discharges) are above average. Since a typical river receives inputs from many tributaries covering its whole drainage basin, for a serious natural flood to occur a significant rise in overall discharge is required and so there must be relatively large increases in inputs from many tributaries. This will happen when a large part of a catchment is subjected to prolonged, high intensity rainfall, but even then the size of water inputs will depend on how saturated the soil on catchments has become due to previous rainfall events.

Although rainfall in the permanently humid tropics is distributed fairly uniformly throughout the year, river discharges fluctuate a great deal. The Congo has two peaks in November–January and April–June, for example, because it receives inputs from tributaries rising in seasonal areas to the north and south of the Equator where rainfall peaks at different times of the year. In monsoonal areas, such as India, where rainfall seasonality is very pronounced even forested catchments have only a limited ability to temper the impact of very intense rainstorms, so flooding is likely there for entirely natural reasons. The major flood of the Ganges and Brahmaputra in September 1988, for example, which affected two thirds of

Bangladesh, was shown to be mainly due to heavy monsoon rainfall rather than deforestation (Ward and Robinson, 1990).

Deforestation of catchments does cause localized flooding of small rivers but there is little evidence that it has yet contributed to the flooding of major rivers such as the Amazon or the Ganges. This would need an increase in flood frequency at the same time as pronounced catchment deforestation and no change in rainfall. Gentry and Lopez-Parodi (1980) thought they had shown such a link between flooding and deforestation in Peruvian Amazonia when they found that the high water level of the River Marañon, an Amazon tributary, rose significantly by 8 per cent between 1962–69 and 1970–78 while rainfall only increased slightly. But Nordin and Meade (1982) argued that this conclusion was biased by the period chosen for the test, so that if 1942–56 to 1970–80 had been chosen instead, for example, there would have been no statistically significant increase in high water level and hence no link with deforestation.

This does not mean there is no need to conserve forest cover on catchments, as this is vital to maintain stable water flows, merely that the role of deforestation in flooding should not be exaggerated. Good catchment management will become even more important in future if, as expected, the enhanced greenhouse effect leads to a wetter as well as a warmer planet (Wigley and Jones, 1985).

Deforestation and sedimentation

There is less doubt about the link between deforestation and higher river sediment loads arising from increased soil erosion. A change from forest to rice cultivation in Peninsular Malaysia, for example, caused a twentyfold increase in stream sediment concentration (Douglas, 1967). But the impact varies from place to place depending on the background erosion rate. Soil erosion occurs all the time for natural reasons and newly uplifted areas, such as the Himalayas, are highly erodible. The suspended sediment load of the River Ganges, 1568 tonnes per sq km per annum, makes that of the Amazon, at 67 tonnes per sq km per annum, look minuscule, even though the Amazon has the largest discharge of any river.

High sediment loads give rise to many problems. Particularly important is the sedimentation of reservoirs having vital roles in hydroelectric or irrigation schemes. The lifespans of the Binga and Ambuklao dams in the north of the Philippines island of Luzon, for example, were cut by a half owing to the effects of sediment build up caused by deforestation. Irrigation canals also silt up: intensive rice cultivation on the Indonesian island of Java has suffered for decades from siltation due to deforestation of upland catchments such as that of the River Solo. The costs of the off-site effects of soil erosion on Java's irrigation systems, harbours and reservoirs have been estimated at \$25–91 million per annum (Magrath and Arens, 1987). Sediment accumulation on the beds of rivers can also raise them above the level of the surrounding land and this exacerbates flooding.

Deforestation and climate change

It is often popularly assumed that tropical deforestation is followed by a general decline in rainfall in the deforested area, but the evidence for this is more anecdotal than based on reliable empirical data so the effect remains in doubt. What is more certain, however, is that carbon dioxide emitted after deforestation is contributing to the general rise in the atmospheric concentration of that gas and this is expected to cause a change in global climate through the greenhouse effect. This is the principal effect of tropical deforestation on the atmosphere: no major impact on atmospheric oxygen is expected.

The possibilities for local climate change

There is no firm evidence to link deforestation in the humid tropics with a subsequent decline in rainfall, or any other change in local climate for that matter, but this does not mean it could not happen, and indeed four mechanisms could be involved. The first is due to an increase in the proportion of incident solar radiation reflected from the Earth's surface, called the albedo. Converting tropical rain forest to cropland or pasture, for example, reduces the density of vegetation cover and raises the albedo from about 13 per cent to 20 per cent (Pinker et al, 1980; Henderson-Sellers and Robinson, 1986). The more solar radiation is reflected the less the atmosphere is warmed so deforestation should exert a cooling effect on the surface. (Solar radiation is rich in the short wavelength visible and ultraviolet parts of the spectrum but after absorption by the Earth's surface it is re-radiated as high wavelength infra-red (heat) radiation which warms the atmosphere). This could have climatic effects but there is no consensus on what they will be. One explanation for the prolonged drought in the semi-arid Sahel region of West Africa, for example, was the theory of biogeophysical feedback, in which a higher albedo, caused by sparse vegetation cover in dry conditions, changes the energy balance of the atmosphere, reduces rainfall, and leads to lower vegetation growth that reduces rainfall even more, perpetuating dry conditions (Charney, 1974, 1975). If this mechanism were to begin to operate in the humid tropics the consequences could be equally serious.

Deforestation should also reduce the large proportion of incident rainfall normally returned to the atmosphere by evapotranspiration from tropical rain forest. In Amazonia, where half of all rainfall comes from water previously emitted by evapotranspiration, deforestation on a large scale could make the regional climate drier (Marques et al, 1980a, 1980b) but it is unclear how large a clearance is needed to have a significant effect. Evapotranspiration, however, is also an important means of transferring energy from the Earth's surface to the atmosphere: energy is needed to vaporize water given out by evapotranspiration, and later released when the water vapour is condensed higher in the atmosphere. By reducing the rate of evapotranspiration deforestation should therefore exert a net warming effect on the surface, opposing the cooling effect caused by a higher albedo.

Deforestation is also likely to cause changes in air flow patterns. Tropical rain forest has quite a jagged profile because of the emergent trees which penetrate the continuous canopy. Air flows over the canopy are therefore suitably erratic. Deforestation leads to a smoother 'canopy' of bare ground, field crops, or tree crop plantations that changes the flow of winds in the area and could affect rainfall patterns (Brunig, 1978).

Finally, on the higher reaches of mountains, above the cloud line, upper montane forests are steeped in fog, some of which condenses on the vegetation and flows to the ground, accounting for a large proportion of precipitation at some times of the year (Whitmore, 1990). If the upper montane forest, often called 'cloud forest', were removed this precipitation would be lost and the immediate area would become drier.

The most convincing conclusion from the limited amount of empirical data on this subject is that while deforestation will not reduce the annual amount of rainfall in an area it could cut the number of rainy days and increase the intensity of rainfall (Meyer-Homji, 1988). The results of regional-scale experiments with general circulation models (see below) that take into account the first three of the above mechanisms are ambiguous: some suggest an increase in rainfall if all tropical rain forest in Amazonia were cleared while others predict a decrease (Henderson-Sellers and Gornitz, 1984; Dickinson and Henderson-Sellers, 1988). Until more data are available we cannot be more definite about the likely impact of deforestation on local and regional climates.

Deforestation and oxygen

Despite popular concern that deforestation is reducing the oxygen content of the atmosphere this has not been regarded as a serious threat by the scientific community for some time. Undisturbed tropical rain forests are climax ecosystems, and so are in a state of balance with the environment, which means that their total biomass stays constant and they only release as much oxygen in photosynthesis as they use up in respiration. For this reason, clearing tropical rain forest should not adversely affect the net annual supply of oxygen to the atmosphere. (Naturally, if the forests cleared are still maturing after a previous clearance there will be an impact.) Another possibility is that burning forests after deforestation would consume an inordinate amount of oxygen and reduce its overall level in the atmosphere. But even if all tropical forests were cleared and fully burnt so their carbon was turned into carbon dioxide the oxygen content of the atmosphere should not fall by more than 0.05 per cent, so there seems no cause for concern here either (UNESCO/UNEP/FAO, 1978; Grainger, 1980a).

Tropical deforestation and the greenhouse effect

The main effect of tropical deforestation on atmosphere and climate results from the carbon dioxide emitted from the burning of cleared vegetation. Given the massive scale on which tropical deforestation is taking place and the large amount of carbon stored in tropical rain forests these emissions, together with those from fossil fuel combustion, are leading to a significant rise in the overall carbon dioxide content of the atmosphere. Though it is only a trace gas in comparison with oxygen and nitrogen, carbon dioxide is an avid absorber of the infra-red (heat) radiation emitted from the Earth's surface following the absorption of solar energy (see above). A large proportion of this heat is normally lost into outer space but enough remains to leave the planet reasonably warm on average. As the concentration of carbon dioxide rises more of this heat is held in the atmosphere and the Earth becomes even warmer. This, very briefly, is the basis of the greenhouse effect, which is expected to increase average global temperature and cause other climatic changes as well (Barry and Chorley, 1987).

A significant rise in the carbon dioxide content of the atmosphere is possible as a result of tropical deforestation as the total amount of carbon currently stored in the atmosphere is similar to that in terrestrial vegetation, and there is twice as much carbon in the soil as well. A large transfer of carbon from biomass and soil to the atmosphere could therefore cause a significant rise in atmospheric carbon dioxide. Tropical rain forests could play a crucial role in this because they contain an estimated 20–40 per cent of all carbon in terrestrial vegetation (Whittaker and Likens, 1975; Brown et al, 1989) (our knowledge of this is still inaccurate). The amount of carbon locked up in fossil fuels is 14 times that in the atmosphere so there is even more concern about the changes that their continuing use could cause.

Carbon dioxide is actually one of a number of gases, which also include methane, nitrous oxide, ozone and water vapour, that absorb infra-red radiation in the atmosphere and prevent it from leaving the Earth. The natural concentrations of these gases in the atmosphere already make the Earth warmer than it otherwise would be. But over the last two hundred years fossil fuel combustion from our industrial activities, continuing deforestation and other human activities have caused their concentrations to rise. These trends, by enhancing the pre-existing greenhouse effect, are turning it from a benefit into a threat. Regular measurements at the Mauna Loa observatory in Hawaii show the carbon dioxide content of the atmosphere has increased by 11 per cent from 315 ppm in 1958 to 349 ppm in 1988 (Figure 6.3) (CDIAC, 1989). In 1700, before the Industrial Revolution began, data from air bubbles trapped in polar ice indicate that the concentration was 275–85 ppm (Neftel et al, 1985; Siegenthaler and Oeschger, 1987).

Table 6.1 *Estimates of carbon dioxide emissions from tropical deforestation (gigatonnes of carbon per annum)*

Source	Year	All tropical forest	Tropical moist forest	Tropical rain forest
Woodwell et al (1978)	1970s			1.0–7.0
Woodwell et al (1983)	1980	1.3–4.2		
Houghton et al (1985)	1980	0.9–2.5		
Molofsky et al (1984)	1980	0.6–1.1		
Detwiler and Hall (1988)	1980	0.4–1.6		
Grainger (1990b)	1980	0.5–0.8	0.4–0.7	

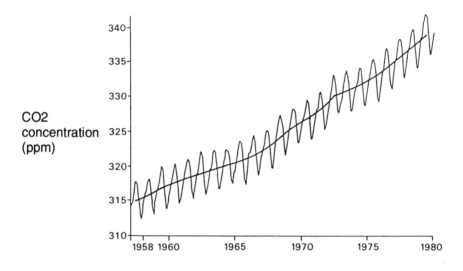

Source: NAS (1983)

Figure 6.3 *Trends in atmospheric carbon dioxide content 1958–80*

The volume of carbon dioxide emissions from tropical deforestation is still uncertain. Estimates have declined since 1978 when George Woodwell (then at the Marine Biological Laboratory in Woods Hole, Massachusetts) and colleagues proposed a figure of 1–7 gigatonnes[1] of carbon (gt C) per annum (Table 6.1) (Woodwell et al,1978). The Woods Hole team later produced an estimate of 0.9–2.5 gt C per annum for 1980 (Houghton et al, 1985) but Paul Detwiler and Charles Hall from Cornell University subsequently gave an even lower estimate of 0.4–1.6 gt per annum for the same year using an improved methodology (Detwiler and Hall, 1988). Such volumes are clearly smaller than the 5.6 gt C per annum emitted in 1986 by fossil fuel combustion, which accounts for the majority of all carbon dioxide emissions from human activities.

It would seem that based on the upper and lower figures from the two 1980 estimates tropical deforestation accounts for between 7 and 30 per cent of all carbon dioxide emissions. The large range is due mainly to inaccuracies in estimates of (a) deforestation rates and (b) the amount of biomass per unit area of cleared forest, though our knowledge of all the processes involved is also quite poor. For example, carbon dioxide is also emitted by oxidation of soil beneath cleared forest, and emissions from this source and burnt vegetation can be delayed for some years after deforestation and burning. Carbon is taken in again by the soil and vegetation when agricultural land is abandoned and properly accounting for this also causes problems.

Deforestation affects the concentrations of other greenhouse gases as well. Methane, nitrous oxide and ozone are all produced when vegetation is burnt; more methane is emitted after deforestation by rice paddies and the guts of cattle raised on ranches; and deforestation causes a drop in the natural production of nitrous oxide from forest soils. Methane has an even greater warming effect than carbon dioxide and its concentration increased at 1 per cent per annum during the 1980s to reach 1.7 ppm – more than double that in the 17th century (Khalil and Rasmussen, 1989). The nitrous oxide concentration has increased more slowly at just 0.2 per cent per annum, reaching 0.33 ppm (MacElroy and Wofsy, 1986; Mooney et al, 1987). How much of these increases is due to deforestation is uncertain.

How will global climate change in future?

To find out how global climate could possibly be changed by the enhanced greenhouse effect we rely on experiments with large computer models, called general circulation models or GCMs, which simulate the many processes of atmospheric circulation. The results of these experiments differ from one model to another, but overall they predict an increase in average global temperature of between 2°C and 5°C should carbon dioxide content rise to twice its pre-industrial (17th century) level of 280 ppm (Schlesinger, 1989).

Although carbon dioxide is only one of a number of greenhouse gases, for convenience climate modellers group them together and model their warming effect in terms of carbon dioxide equivalents. So while at the present rate of increase the actual level of carbon dioxide in the atmosphere should not double by 2030, the overall warming effect could be equivalent to this if a rise to 433 ppm, for example, were accompanied by simultaneous concentration increases of 64 per cent in methane, 16 per cent in nitrous oxide, 13 per cent in ozone and 600 per cent in chlorofluorocarbons (Mintzer, 1987).

Deforestation and industrial activities have been taking place for some considerable time, so we should have already experienced some warming from the carbon dioxide emissions associated with them. GCM experiments have been undertaken to tackle this question too and have predicted, again with much variation between models, that a warming of 0.3 – 1.1°C should have occurred between 1850 and 1980 due to emissions over that period (Wigley and Schlesinger, 1985). This provides a basis for testing the validity of greenhouse effect theory.

There is indeed evidence of an overall rise in average global temperature of 0.5°C between 1900 and 1984 (Figure 6.4) (Jones et al, 1986), but this included a prolonged period between 1940 and the mid-1960s when temperatures actually fell and this needs explanation. Some experts have blamed the effects of volcanic eruptions. Others have claimed that observed changes are well within the bounds of natural variation (Schneider and Rosenberg, 1989) and that the existence of the enhanced greenhouse effect is still open to doubt.

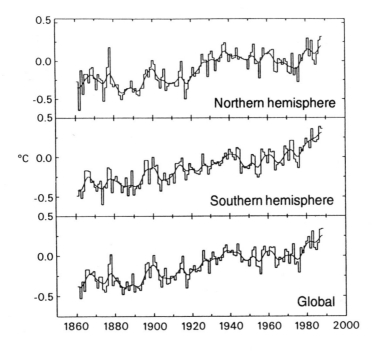

NB. The trend in average surface air temperature is expressed relative to the long-term mean.
Source: Reproduced by Grainger (1990a), redrawn from Jones et al (1988), updated by the original authors

Figure 6.4 *Trends in average global, northern hemisphere and southern hemisphere surface air temperatures, 1901–88*

Even if the effect is operating it is still difficult to predict how global temperature is likely to rise in future. We do not know how fast the concentrations of carbon dioxide and the other greenhouse gases will rise in future as this all depends on trends in world patterns of energy consumption and land use and other unknown factors. For example, the way that all the carbon dioxide we emit is stored or cycled around the planet is not fully understood, and there are numerous delays in the response of the atmosphere to our emissions, owing to the need to establish gaseous equilibrium between the atmosphere and the oceans and between different ocean layers. This, and the equally important delay in establishing thermal equilibrium between the atmosphere and the oceans, lead to a considerable lag in global warming, so it could take 20 to 30 years for the effect of any carbon dioxide emitted now to become fully apparent. According to the World Meteorological Organization, by the year 2030 we shall have committed ourselves to a rise in average global temperature of 0.8 to 4.1°C but only half of that warming will have occurred by then (WMO, 1986).

Global climate models suggest the warming trend will not be felt uniformly across the planet. Areas near the Equator should experience the smallest rise in temperature and areas near the Poles the highest. Changes in precipitation patterns are predicted as well as a warming trend, as higher temperatures mean more evaporation of water into the atmosphere and a corresponding rise in rainfall and other precipitation. Unfortunately, different GCMs lead to conflicting conclusions about whether rainfall in different parts of the humid tropics will increase or decrease, and how its distribution throughout the year will change. Because tropical rain forests crucially depend not only on high rainfall but also minimal seasonality in its distribution throughout the year we are therefore poorly placed at the moment to assess the impact that global climate change will have on the humid tropics.

Impacts of climate change on agriculture and the environment

Bearing in mind these limitations we now review briefly some of the possible impacts on tropical agriculture and environments. The growth of crops and trees will be affected by their physiological response to the sum of changes in temperature, rainfall and carbon dioxide. For if temperature rises and rainfall stays the same this would reduce the availability of water to the crop and yields will decline. Carbon dioxide acts like a fertilizer, but its impacts vary from one crop to another: some crops, such as rice and wheat, are liable to increase their growth rates as the carbon dioxide level of the atmosphere rises. Others, such as maize, which have a different form of photosynthetic pathway, will show only a minor yield increment due to this reason (Schneider and Rosenberg, 1989). To these effects on yields we must add that due to any change in the incidence of pests and diseases as a result of climate change.

Considerable effort has been expended in studying the possible impact of global climate change on world agriculture (Parry et al, 1988; Parry and Carter, 1989) but no firm conclusions have yet been reached about the tropics because of uncertainty, mainly about changes in rainfall. According to a study by the Environmental Change Unit at Oxford University higher temperatures could reduce rice yields in South-east Asia but these could be offset by the effects of more rainfall (ECU, 1990). Agriculture will also be affected as rising sea levels, caused by the melting of mountain glaciers and polar ice sheets and thermal expansion of oceans, put low-lying farmland near coasts and river estuaries at greater risk from flooding and saline intrusion. An average global rise in sea-levels of 0.2–1.65 m by the year 2100 has been predicted (Robin, 1986). This could flood as much as a third of Bangladesh, together with up to 40 per cent of Indonesia, the coastal plain of Guyana, and coral island nations in the Pacific and Indian Oceans such as the Maldives (Mdoud, 1989; Hekstra, 1989; Sattaur, 1990; Wells and Edwards, 1989).

Changes in forest growth rates are also likely as a result of changes in temperature, rainfall and carbon dioxide content, but overall they are likely to be greatest at higher latitudes (Sedjo and Solomon, 1989). Another expected consequence of global climate change is a shift in the distribution of the world biome types. For example, in the northern hemisphere temperate deciduous forest should spread into the boreal forest which would in turn move northwards on to pre-viously unforested land (Parry and Carter, 1989). The extent of any shift in tropical rain forest will depend on how much seasonality changes, and the forest's speed of response to any climate shift will be hampered by the slow pace at which primary forest trees can colonize new areas and by barriers imposed on the spread of natural vegetation by farmland. Meanwhile, tropical rain forest would come under new threats as greater climatic variability caused by the green-house effect increased the frequency of droughts and forest fires. Some idea of what could happen may be gained from the 3.5 million ha of tropical rain forest in East Kalimantan damaged by drought and fire in 1982–3 (Lennertz and Panzer, 1983). The problem recurred in 1991 (Schwarz, 1991).

Tropical rain forests and climate change mitigation

The world community is now considering how to mitigate the scale of future global climate change and the serious effects that could result from it. The scope for taking action was discussed as part of the deliberations of the Intergovernmental Panel on Climate Change, set up by the United Nations to prepare the scientific background for the Second World Climate Conference, held in Geneva in November 1990 (Houghton et al, 1990). Work then began to draft the Climate Change Convention for signature at the UN Conference on Environment and Development in Rio de Janeiro in June 1992 (see Chapters 7 and 8). Although the Convention did not, as expected, commit signatory countries to making any specific reductions in their carbon dioxide emissions, it is hoped that this will happen in due course.

The main focus of future action to reduce carbon dioxide emissions is likely to involve curbing fossil fuel use but there is also scope for action in the forest sector. One way to do this is by controlling tropical deforestation, the second most important source of emissions. Based on alternative scenarios for controlling deforestation by improving farm productivity (see Table 9.3), this author has suggested that carbon dioxide emissions from deforestation in the dry and humid tropics could fall to about 0.2–0.5 gt C per annum by 2020, compared with 0.5–0.8 gt C per annum in 1980 (Table 6.2) (Grainger, 1990b)[2].

Table 6.2. *Projections of possible future trends in carbon dioxide emissions from deforestation of tropical moist forest and all tropical forest (gigatonnes of carbon per annum)*

	High scenario					Low scenario				
	1980	1990	2000	2010	2020	1980	1990	2000	2010	2020
Humid tropics	0.702	0.661	0.572	0.495	0.434	0.449	0.391	0.291	0.205	0.126
All tropics	0.797	0.757	0.668	0.591	0.530	0.544	0.486	0.387	0.300	0.221

NB. Simulations for all tropical forest include the following constant values of emissions from the dry tropics: Africa 0.043, Asia 0.009, Latin America 0.044, total dry tropics 0.096 (gt C). (See Table 9.3 for the original deforestation scenarios.)
Source: Grainger (1990b), extending deforestation modelling work described in Grainger (1990d)

Another way is to increase the rate of carbon uptake by tropical forests. Already, tropical rain forests regenerating after clearance or logging accumulate new biomass and so constitute a major carbon sink. Protecting such forest will therefore help to offset net carbon emissions from other sources. But the size of the sink could be increased dramatically by planting a huge new area of forest (Sedjo and Solomon, 1989). To fully absorb present carbon dioxide emissions from all sources would require almost 600 million ha of forest if it were sited on the tropics, where growth rates are high and a large area of degraded land is available (Grainger, 1988b). But this would involve expanding the area of all tropical closed forest in 1980 by an extra 50 per cent, so the feasibility of such a massive project must be doubtful. Planting the new 'carbonforest' over 20 years, for example, would require a 30-fold increase in the current planting rate (Grainger, 1990c). There would also be a host of problems, including what to do with all the wood produced, and the social and political difficulties arising from the displacement of people from the land needed for planting. The governments of tropical countries might also not be too happy if they were asked to bear the brunt of the world's fight to mitigate global climate change, even if they were financially compensated for doing so.

It would seem more realistic to aim for a more modest increase in the planting rate and only expect tropical countries to reabsorb some of the carbon dioxide that they have emitted themselves. Combining a programme to control deforestation with a tripling in the current planting rate to 3 million ha per annum could lead to zero net emissions from deforestation by the year 2020 (Table 6.3) (Grainger, 1990c). This would seem feasible both economically and politically, for the tropical countries would now be called on only to absorb their own emissions, and not those from the industrialized countries as well.

The likely consequences of tropical deforestation

Tropical deforestation is having, and will continue to have, significant environmental impacts. Biological diversity is being reduced by deforestation and logging, with effects ranging from an erosion of the genetic diversity of some species to the extinction of others. The tragedy of extinction is that we lose some species before we even know they existed. Less extreme impacts may prejudice our ability to breed new varieties of crops to cope with new pests and diseases. The high concentration of species (half of the world total) in the tropical rain forests is one of the greatest justifications for conserving them.

Deforestation and logging are also causing soil degradation, changes in water flows and increased sedimentation of rivers, reservoirs and irrigation systems. The scale of these impacts can be significantly reduced by better forest management, choosing the

Table 6.3. *Alternative scenarios for a carbonforest programme*

A. Afforestation scenarios for a carbonforest to achieve a range of reductions in annual atmospheric carbon dioxide increment over periods of 10, 20, 30 and 50 years.

Percentage reduction in annual CO_2 increase	Total planted area (million ha)	Target planting rates (million ha/yr)			
		10	20	30	50 (yr)
Base scenario					
100	600	60	30	20	12
50	300	30	15	10	6
25	150	15	8	5	3
10	60	6	3	2	1

B. Integrated afforestation and deforestation-control strategies. Net carbon emissions resulting from the combination of carbon emissions by all tropical deforestation (High and Low scenarios) and carbon uptake by the carbonforest (various planting rate scenarios)

Planting rate (million ha/yr)	Net carbon emissions (gt/yr)									
	High deforestation scenario					Low deforestation scenario				
	1980	1990	2000	2010	2020	1980	1990	2000	2010	2020
12	0.757	0.699	0.030	−0.628	−1.269	0.544	0.486	−0.152	−0.832	−1.499
6	0.757	0.728	0.349	−0.018	−0.370	0.544	0.515	0.167	−0.223	−0.599
3	0.757	0.742	0.508	0.286	0.080	0.544	0.530	0.327	0.082	−0.150
2	0.757	0.747	0.562	0.388	0.230	0.544	0.535	0.380	0.184	0.000
1	0.757	0.752	0.615	0.489	0.380	0.544	0.539	0.433	0.285	0.150

NB. The net carbon uptake by the 7.1 million ha of existing tropical industrial wood plantations in 1980 adds an estimated 0.034 gt per annum to the above.
Source: Grainger (1990c)

land uses introduced to particular sites more carefully, and being firm in protecting catchment forests from clearance and intensive logging.

Tropical deforestation is recognized as making a major contribution to potential global climate change resulting from the enhanced greenhouse effect, though there is less certainty about its possible impact on local and regional climates. Tropical countries will suffer from future climate change just as much as temperate countries but the role of tropical rain forest as a global carbon sink and store is so important that as part of a future global initiative to mitigate climate change they could receive substantial financial aid to protect their forests and increase the rate of tropical reforestation. However, the scale of such action will be limited by the same social, economic and political constraints that keep the present planting rate in the whole tropics down to only 1 million ha per annum.

7

Techniques to control deforestation

It is vital to bring the deforestation of tropical rain forests under control, but how can this be done? Unfortunately, there are no quick solutions and it is unlikely that tropical rain forest management can be made fully sustainable quickly either. A long-term programme of action is required to tackle both problems over the next 10 to 20 years, and the longer we wait before starting the longer it will take. The more time is lost on attempting facile 'quick fixes' the more money will be wasted and the greater the loss of forest.

If tropical deforestation is an inevitable consequence of social and economic development, as this book has argued, it is unrealistic to think of halting it overnight. But it can be controlled by raising farm productivity and sustainability and concentrating the most intensive practices on the most fertile lands. Forest management also takes time to evolve, but the process could be speeded up by increasing technical and financial assistance to tropical countries, and significant short-term improvements could be gained by better monitoring and regulation of logging practices and more effective protection against illegal clearance.

Of course, the ideal would be to preserve a large proportion of the remaining tropical rain forest in national parks and other reserves as soon as possible, free from all future human interference. But increasing the amount of tropical rain forest in protected areas will be a slow process too. So it is important in the short term to focus on safeguarding those areas of forest having the highest priority for conservation, and ensure that the boundaries of these and existing protected areas are respected and deforestation does not spread over them. To save as much of the forest as possible in the long term, conservation must be combined with actions to reduce the pressures leading to deforestation. This can be achieved by the initiatives in the agriculture and forest sectors suggested here, and also by more intelligent planning so that, for example, the building of roads through pristine forested areas is kept to a minimum. Such an approach should give conservation workers time to do their job properly.

There are three keys to controlling deforestation. The first is to take an integrated approach that embraces agriculture, forestry and conservation. Previous schemes failed in part because they included only forestry and/or conservation components and usually ignored agriculture entirely. Second, the underlying causes of poor land use must be tackled as well as land management practices. Deforestation is a symptom of wider social and economic conditions, and although poverty and inequality in developing countries cannot be changed overnight it is important to address these underlying causes so that the right balance can be struck between conservation and the needs of development. Third, better techniques by themselves are not sufficient and they need the support of sound government policies. Changes in policy are necessary also to deal with the underlying social and economic causes of deforestation and to correct former inappropriate policies that have contributed significantly to the problem. On the other hand, deforestation is not caused only by poor government policies, as some would argue, and controlling it requires a judicious mix of improved management techniques and policies that balance local and centralized action.

This chapter and the next outline an integrated approach to the tropical rain forest problem which simultaneously:

- reduces the rate at which agriculture expands by promoting higher farm productivity and sustainability;
- makes forest management more sustainable and improves forest protection against clearance;
- increases the amount of forest conserved in protected areas.

The main recommendations are summarized in Table 7.1. The emphasis in this chapter is on better land management techniques, illustrated by examples wherever possible, which can be promoted by governments, international development agencies and non-governmental organizations. Recommendations for policy changes are made in the next chapter.

Improving farming practices

Since agricultural expansion is the main reason for deforestation, improving farming practices should be at the core of an integrated control programme. Of the wide variety of land uses involved most need some modification to become more sustainable and productive, but the overall philosophy behind the strategy outlined here is

Table 7.1. *Techniques for controlling deforestation*

A. Make agriculture more productive and sustainable
1. Improve land use policy and planning
2. Improve shifting cultivation
   a. Make short-rotation shifting cultivation more sustainable
   b. Control spread of encroaching cultivation
   c. Introduce appropriate agroforestry techniques
3. Improve permanent agriculture
   a. Promote permanent cultivation on good sites but use incentives more carefully in future
   b. Make resettlement programmes more sustainable and productive
   c. Curtail spread of cattle ranching
   d. Promote tree crop plantations
   e. Introduce appropriate agroforestry techniques

B. Make forestry more productive and sustainable
1. Place new emphasis on natural forest management
2. Improve implementation of current selective logging systems
   a. Make monitoring and regulation more effective
   b. Give incentive to concessionaires by increasing concession periods
3. Increase protection of logged and unlogged forest
4. Improve selective logging systems
5. Develop new silvicultural systems
6. Continue establishing timber plantations

C. Conserve more forest within protected areas
1. Promote passive conservation: wise planning to reduce deforestation
2. Choose protected areas to maximize conservation of national/ international biodiversity
3. Formulate national conservation strategy
4. Design protected areas for long-term viability
5. Pay greater attention to local participation and buffer zones
6. Devise new conservation financing mechanisms

---

to build on existing practices rather than to replace them by entirely new ones. This is followed in every case except that of encroaching cultivation, where the solution is more social and political than technical. Most of the proposed changes can be made by farmers themselves, without central direction or a massive change in social systems.

This is not to deny the potential for devising new farming systems that are more appropriate to environmental conditions in the humid tropics, though getting them adopted by farmers will not be easy. However, there is great scope to introduce agroforestry systems that combine trees with field crops and/or livestock raising, and agroforestry can often be introduced in ways that enhance existing practices. This section recommends priority actions for each major land use type in turn, selected from a range of possible actions on the basis of wide applicability. Other solutions have undoubtedly worked in certain areas but have only a limited appeal on the larger scale and so are not mentioned here.

Take no action on traditional shifting cultivation

Traditional, long-rotation shifting cultivation is one of the few proven land uses in the humid tropics but because of its extensive nature and comparative rarity these days it does not represent a threat to tropical rain forests or warrant urgent remedial action.

Improve short-rotation shifting cultivation

Most shifting cultivation is now practised on shorter rotations. Farmers usually have the intention of sustainability but as the fallow period gets shorter this aim is placed increasingly under threat owing to increased soil erosion and nutrient depletion, greater weediness and falling yields. Because it results in large areas of contiguous forest fallow and cultivation the impact on forest is significant and it is essential that it become more sustainable so the rate of spread into uncleared forest is curtailed. Owing to the large number of people who practise it, outright replacement is not feasible. Nor would it be desirable, since shifting cultivation may well be the most appropriate land use for large areas of low fertility soils. What is needed, therefore, is to improve productivity and sustainability by building on existing practices in ways that make them even more attractive to farmers.

Could fertilizers halt nutrient decline?

The sustainability of short-rotation shifting cultivation is threatened by the decline in fertility as fallow periods are reduced. So from a purely technical perspective it might be appealing to persuade farmers to use fertilizers to compensate for this. Trials on ultisol plots at the Yurimaguas Experiment Station in Peru, led by Pedro Sanchez of North Carolina State University, showed that this was feasible. Applying lime and fertilizers containing nitrogen, phosphorus, potassium, molybdenum and boron to land cleared from 17-year old secondary forest lowered soil acidity and exchangeable aluminium concentrations, increased the concentration of major soil nutrients, and gave yields of 2.71 and 2.81 tonnes per ha of rice and maize respectively in the eighth year of continuous cropping, compared with only 0.99 and 0.21 tonnes per ha on an unfertilized plot (Sanchez et al, 1983). However, the high cost of fertilizers and the need to monitor soil conditions closely make this innovation inappropriate for most shifting cultivators, so it is of little practical relevance at the moment.

Introduce new low-cost systems

Appreciating these problems, Sanchez and his team tested another approach that required only low-cost inputs. No fertilizers were used, the land was not tilled, crop residues were returned to the plot after harvesting, acid-tolerant cultivators of rice and cowpea were grown and weeds cleared regularly. After three years of cultivation and seven crop cycles yields were generally equivalent to those on fertilized plots and the system was more profitable overall because costly fertilizers were avoided, although the land would then need weeding and fallowing before more crops could be grown (Sanchez and Benites, 1987). The system has promise but requires further testing under more realistic conditions to confirm its wider applicability. Results so far show that short-rotation shifting cultivation can be improved even if only low cost inputs are possible. But to encourage farmers to enhance their practices governments should appoint more agricultural extension officers who can visit farmers regularly and advise them on this and other improved techniques.

Restore fertility with planted tree fallows

Instead of relying on natural forest regeneration to restore fertility, shifting cultivators could plant nitrogen-fixing leguminous trees, such as Gliricidia sepium and Leucaena leucocephala, to speed recovery of soil nutrients. The trees could be also cropped for food, fodder, fuelwood and other products and this would improve the productivity of the system too. This is not a new approach and is already found in various countries. In southwest Nigeria farmers use the green stems of Gliricidia sepium as stakes for yams. After the yams have been harvested the gliricidia trees, which are a valuable source of fodder and green manure, are left to carry on growing during the fallow period. They are cut down when it is time to plant crops again but will subsequently regenerate from the stools that remain (Getahun et al, 1982).

The taungya system

In forests dedicated to long-term timber production food crops can be interplanted with tree seedlings on newly cleared lands for two to three years until the trees are large enough for the canopy to close. This is the basis of the taungya system devised by a Burmese forester, U Pan Hle, in 1856. He employed local farmers to plant teak seedlings and in return allowed them to plant food crops for themselves as well. Taungya later spread to India, Thailand, Nigeria and Sierra Leone and other countries. Now applauded for integrating forestry and agriculture, it was for a long time simply a cheap way for foresters to establish timber plantations and alleviate the pressures for deforestation, the farmers being employees rather than partners in the operation (Blanford, 1958; King, 1989).

A general lack of attention to farmers' needs is probably one reason why taungya has not stopped the spread of shifting cultivation where it is practised. The government of Thailand tried to tackle this obstacle in 1967 by starting a voluntary 'forest villages' scheme in which shifting cultivators were paid to reforest a target area of 1.6 ha each year, given a plot of land in specially planned forest villages on which to build a house and establish a multi-storey homegarden (see below), and received free medical care, education, electricity, drinking water and other benefits. However, the scheme has not been successful, for only 26 out of the intended 2000 villages had been established by 1981. It was hard to persuade shifting cultivators to settle down, and they were also unhappy with the money they received (Boonkird et al, 1984).

Agroforestry systems
Planted tree fallows and taungya are examples of an agroforestry system called agrisilviculture in which trees and agricultural crops are grown together in rotational or spatial mixtures. Many of the new farming practices developed for the humid tropics are likely to involve trees in some way, usually in a 'multi-storey' system that mirrors the structure of tropical rain forest itself. Agroforestry is the collective name for a variety of practices that combine the growing of trees with field crops and/or the raising of livestock, and has been defined more formally as:

> The collective name for a group of land use systems in which woody perennials, including trees, shrubs, palms, bamboos, etc., are deliberately used on the same land management units as agricultural crops and/or animals, in some form of spatial arrangement or temporal sequence, and with ecological and economic interactions between the different components of the systems (Lundgren and Raintree, 1982).

A number of agroforestry systems are mentioned in this chapter.

They are conveniently divided into four main groups: agrisilviculture, silvopasture, agrosilvopasture and other systems (Table 7.2). Agrisilviculture is the largest of these and also includes traditional long-rotation shifting cultivation, multi-storeyed tree gardens, alley cropping and intercropping in tree crop plantations (Nair, 1989a).

Table 7.2. *Types of agroforestry systems*

Type	Characteristics	Examples
Agrisilviculture	Combines trees and agricultural crops in rotational or spatial mixtures	Traditional long-rotation shifting cultivation; planted tree fallows; taungya system; tree gardens; alley cropping; intercropping in tree crop plantations.
Silvopasture	Combines trees, pastures and animals	Grazing animals under plantations of tree crops, or trees grown for timber, fuelwood or fodder; growing fodder trees in small plantations, or as hedges around pastures.
Agrosilvopasture	Combines trees, agricultural crops pastures and animals.	Growing fodder trees as hedges around fields of crops; mixed raising of trees, livestock and crops around homesteads; other integrated production systems.
Other systems		Mixed plantations of multi-purpose trees with two or more of: food, fodder, fuelwood, timber, extractives; apiculture (mixed bees and trees); aquasilviculture (mixed fish and trees).

Source: based on Nair (1989b)

Agroforestry gardens
The principles behind taungya are fine but social and economic factors have prevented it from being successful. The Nigerian forester Philip Kio has suggested that the way out of this dilemma is for small farmers to become independent timber growers rather than mere labourers employed by forestry departments (Kio and Ekwebalan, 1987) and this makes a lot of sense. One option is to designate distinct

'agroforestry zones', situated between government forest reserves and local villages and their associated fields, in which farmers would establish 'agroforestry gardens' consisting of mixtures of commercial and subsistence tree crops and managed timber trees. The gardens should be profitable enterprises for the farmers and also act as 'buffer zones' between forest and totally cleared areas, reducing the risk of further deforestation. Agroforestry gardens already exist on the Indonesian island of Sumatra, where small tree crops such as coffee are intercropped with larger fruit trees like durian and commercial timber trees, such as *Toona sinensis* and *Pterospermum javanicum*, which are sold after harvesting. A group of people from the local community are responsible for managing the gardens, each member of the group having rights to a 0.63 ha plot (Michon et al, 1986).

Tree gardens
Another type of multi-storey cropping is the tree garden, called a 'homegarden' on the Indonesian island of Java. Tree gardens are planted in or close to house compounds with the emphasis on growing subsistence crops for domestic use. They could be used to improve the food supplies of short-rotation shifting cultivators who are settled in villages, and are also suitable for use in resettlement programmes. Javanese homegardens contain up to 1700 plants per ha, which are cropped for food, spices, medicines, fuelwood, etc. (Terra, 1954; Stoler, 1975). Tree gardens elsewhere differ slightly from those on Java, some containing mixtures of commercial and subsistence tree crops, with the occasional use of hired labour.

Control the spread of encroaching cultivation

Curbing encroaching cultivation is essential but cannot be achieved simply by better farming techniques. Since encroaching cultivation is a symptom of poverty and inequality and practised mainly by landless people, to solve the problem governments must modify their social and economic policies to prevent people having to migrate in the first place. This requires land tenure reform, increased aid to overcrowded areas, and the creation of more jobs in urban areas to reduce urban–rural migration. Road construction through forested areas should be minimized and those roads that are built should have buffer zones of plantations on either side to reduce access to undisturbed forest. Existing encroaching cultivators could be invited to join government resettlement schemes, paid to reforest or otherwise revegetate degraded lands, or be given technical advice by agricultural extension officers if the land they are now using has potential for sustained cultivation.

Promote permanent staple crop cultivation on good sites

The cultivation of staple field crops should be encouraged where the land is suitable, which is the case for much existing wet-rice cultivation, for example. Farmers already settled on sites inappropriate for intensive cultivation should be encouraged, where possible, to adopt improved techniques to increase sustainability, e.g. using terracing to reduce soil erosion on land with a significant slope, but governments should not promote new deforestation on such lands.

One way to enhance permanent cultivation is to intercrop field crops with soil-improving leguminous trees such as *Leucaena leucocephala*. In an experimental system called alley cropping – yet another form of agrisilviculture – which was devised in the 1970s at the International Institute for Tropical Agriculture in Ibadan, Nigeria, maize is grown in the corridors or 'alleys' between rows of leucaena and other trees, which also form the basis for a planted fallow after the crops are harvested. Productivity and sustainability are both good but maize yields are lower than those of fertilized monocultures (Wilson and Kang, 1980; Getahun et al, 1982). An alternative is to use an agroforestry system called agrosilvopasture in which tree hedges are planted around fields and the trees lopped for fodder. *Leucaena leucocephala* is a popular hedging tree on Java. Foliage pruned from it is fed to animals and spread on the ground as a mulch, increasing soil nitrogen and organic matter as well as being a natural fallow and reducing the risk of soil erosion (Juo and Lal, 1977).

The expansion of fish farming in artificial ponds has led to considerable coastal deforestation in some countries, such as Ecuador, often removing the valuable protective functions of mangrove forests, e.g. in preventing coastal erosion, trapping silt to build new land and protecting low-lying farmland from flooding should sea levels rise. But it is possible to avoid this deforestation by practising intensive fish farming within the mangrove swamps, using the agroforestry system known as aquasilviculture. This improves the productivity of traditional fishing practices while retaining the environmental services of mangrove forests.

It is unrealistic to expect the expansion of permanent staple crop cultivation to end. But governments should ensure that most of it takes place on fertile land, much of which is still unused for farming. They should identify such land by undertaking comprehensive national land suitability assessments, and use these as the basis of an integrated national land use policy, as discussed in Chapter 8.

Governments should also be careful to avoid promoting the spread of permanent agricultural systems that can only be commercially sustainable with large subsidies. Too often in the past government subsidy schemes have distorted normal market controls and led to the spread of inappropriate land uses, extensive deforestation and other forms of environmental degradation.

Curb the spread of cattle ranching

Nowhere is this more true than in Brazilian Amazonia, where the government-promoted expansion of cattle ranching has resulted in the clearance of huge areas of tropical rain forest. Few of the ranches have turned out to be productive or sustainable but Brazilian ranchers have been happy to take advantage of government handouts, while paying much less attention to setting up productive livestock raising enterprises. Ranching could be sustainable on appropriate sites in humid tropical areas – except in areas in Africa afflicted by sleeping sickness disease – but it is clearly wrong to subsidize it (or other land uses) where it is unproductive and unsustainable. All remaining subsidy schemes of this kind should therefore be ended, and fortunately the Brazilian government has already taken this step.

Halting subsidies still leaves the problem of what to do with present ranches. Forest should regenerate naturally on most abandoned land, albeit after some delay, but where livestock are still being raised landowners should be encouraged to increase productivity and sustainability by fertilizing pastures, sowing nitrogen-fixing legumes, establishing timber plantations, or establishing tree crop plantations under which animals can graze – the type of agroforestry called silvopasture, mentioned above.

There is already some experience to draw on when restoring unproductive ranches. Research at the International Centre for Tropical Agriculture in Cali, Colombia, has demonstrated the potential of nitrogen-fixing legumes, such as *Stylosanthes capitata*, to improve pasture productivity (Sanchez, 1982). In the drier north-east region of Brazil, cattle are grazed on either planted pastures or spontaneous grass or shrub growth beneath coconut and carnauba palms (*Copernica prunifera*) and cashew trees (*Anacardium occidentale*) (Johnson and Nair, 1985). A major new initiative is required to test various silvopastoral systems to find those most appropriate for cattle ranches in Brazilian Amazonia.

Make resettlement schemes more sustainable and productive

Planned deforestation, in which governments clear forest for permanent agricultural schemes, should in theory be more sustainable than unplanned deforestation by shifting cultivators since sites can be carefully chosen by governments after their agriculture departments have assessed which areas are suited to continued intensive permanent cultivation. But theory and practice do not always coincide. As Chapter 2 showed, poor choice of sites has led to frequent problems with government resettlement schemes in Indonesia, Brazil and other countries in the past. If such schemes are to continue locations should be selected with more care in future. A number of the farming techniques advocated in this chapter, for example intercropping and tree gardens, could be introduced to improve the productivity and sustainability of farming on existing resettlement projects.

Promote tree crop plantations

Plantations of tree crops such as rubber and oil palm, can, if properly managed, be a sustainable replacement for forest even though they are not equivalent to it in many respects. They should therefore be encouraged on good sites. Corrective measures are seldom needed for existing plantations, but it is important to retain some surface vegetation cover to reduce soil degradation, as long as this does not lead to pest and disease problems.

The owners of existing plantations could enhance them with an agroforestry approach in which the tree crops are interplanted with other food or cash crops. It is quite feasible to increase overall returns without cutting the productivity of the main crop. For example, intercropping coconut with black pepper, cocoa and coffee increased average annual yields from 5172 coconuts per ha to 5466, 6738 and 7318 coconuts per ha, respectively, over a four year trial at the Coconut Research Institute in Sri Lanka. Coconut yields increased for various reasons, one of which was the effects of the fertilizer and manure applied to the secondary crop (Liyanage et al, 1984). There are many possibilities for intercropping – rubber and cocoa is another excellent combination (Alvim and Nair, 1986) – so the final choice depends on site conditions, farmer preference and local commercial considerations.

Tree crop plantations can also be enhanced by silvopasture, a type of agroforestry system in which animals are grazed on pastures established under the tree crops. Combined cattle grazing and coconut cultivation, for example, is common in Papua New Guinea, the Solomon Islands and Fiji and has proved successful (Vergara and Nair, 1985). Well managed pastures also provide a continuous surface vegetation cover, which reduces the risk of soil erosion.

The contribution of agroforestry

A number of the above suggestions for improved farming practices involved agroforestry, but it must not be seen as a universal panacea. Countless 'instant solutions' to problems gained great publicity in the past but eventually met with a sticky end. What agroforestry provides is a framework for improving the productivity and sustainability of existing practices in ways that correspond to the needs of farmers and foresters and, because of the tree component, convey many environmental benefits, such as improved nutrient cycling and reduced soil erosion, as well as a greater resilience to pests and diseases compared with monocultures (Grainger, 1980b, 1990a).

Despite its novelty as an integral discipline there is long experience of what is now called agroforestry. Shifting cultivation has been practised for thousands of years and the taungya system dates back to the middle of the last century. Farmers are often aware of their local agroforestry heritage and so do not see it as an alien introduction. As agroforestry stresses crop combinations farmers can usually continue growing their old crops while introducing new ones. This reduces any actual or perceived risk involved in changing practices and is in line with the philosophy that there is a greater chance of success if changes in farming systems build on present practices, rather than replacing them completely.

Agroforestry does have its limitations, however, which should be appreciated when planning to introduce it. They include competition and other negative interactions between crops, difficulties in mechanization and higher labour costs. The traditional compartmentalization of forestry, crop cultivation and livestock raising, one of the reasons for the problems afflicting environmental management in the tropics today, also makes agroforestry a major challenge to bureaucrats, agronomists and foresters steeped in the old ways.

Research should enable us to avoid some of the present limitations of agroforestry and reduce the significance of others. The body of systematic knowledge is increasing all the time owing to the vigorous efforts of practising farmers and foresters keen to implement systems in the field and scientists studying how to design systems to optimize multiple crop production (Nair, 1989b). While agroforesters are inventing new systems, such as alley cropping, they are also carefully studying existing systems to lay the basis for wider dissemination.

Improving forest management

Managing forests for timber production has proved an excellent way to retain forest all over the world and the humid tropics should be no exception to this. The general aim of management is to maintain forest cover with a good growth rate and an abundance of well-formed commercial timber species so that timber can be harvested on a predictable basis. This requires some human intervention to protect forest, regulate how it is harvested, and balance the drain on timber reserves and incidental damage caused by harvesting with varying degrees of support for, or manipulation of, subsequent forest regeneration.

Historical experience in temperate countries shows a regretable tendency in the initial stages of forest exploitation to 'mine' forests for wood as if they were non-renewable resources rather than to manage them from the start as the renewable resources they really are. Sustainable management is usually only adopted gradually. This pattern is being repeated in the tropical rain forests today, under conditions far harder for management than in temperate forests. So calls by environmentalists to aim for sustainable forest management throughout the humid tropics within five years or so are a bit unrealistic. Nevertheless, with a sound strategy backed by international funding and political will on the part of governments present practices could be greatly improved within a decade or two instead of hundreds of years.

Two main causes of unsustainability were identified in Chapter 3. First, officially recognized selective logging management systems have flaws. Second, the systems are poorly implemented. The immediate aim should therefore be to improve implementation of existing systems, while over the longer term conducting trials and research to devise more sustainable ones. This section suggests how to address these two matters and also discusses whether:

1. low-impact selective logging, which relies on natural forest regeneration, should be replaced by more intensive systems of tropical silviculture in which foresters manipulate regeneration directly;
2. timber production should be concentrated in plantations instead of natural forests;
3. production of timber should be integrated with that of minor forest products.

Improve implementation of selective logging systems

The problems affecting the implementation of present selective logging systems were discussed in detail in Chapter 3. Government regulations on the management of concessions are widely flouted by loggers who take out too much timber per hectare, cause excessive damage to remaining trees needed for the second harvest, build logging roads poorly so they become prone to soil erosion and other problems, take a second harvest before the end of the official harvest cycle, and make little effort to protect forests against illegal clearance by outsiders.

Improve monitoring and regulation

Loggers can get away with these infringements because their activities are not monitored properly on the ground by government forestry officials. The only way to overcome this problem is for the government to hire more forest officers, station them in or near logging concessions, support them with frequent helicopter monitoring and remote-sensing data (see below), and offer salaries high enough to discourage them from accepting bribes from loggers when regulations are broken. When breaches are discovered forestry officials should make sure that concessionaires are punished by fines, or by more extreme measures, such as termination of the concession agreement, if warranted.

Increase concession periods

A 'carrot-and-stick' approach should be more successful than the stick alone, however. Most concessionaires have little incentive to practise sustainable management at the moment as they have no guarantee of recouping any money they invest in long-term measures. Concession agreements usually last for about 20 years compared with a harvest cycle of 30 years or more, and could be cancelled and transferred to others before the second harvest is taken. It is therefore important to extend concession periods to at least 40 years so that one full harvest cycle is covered. Some experts have suggested raising it to 60 to 80 years, i.e. two harvest cycles. Governments would naturally retain the right to cancel agreements at any time if forests were not managed satisfactorily, but concessionaires would at least have an incentive to manage for the future instead of thinking only of the present.

Improve forestry finances

It has also been suggested by some economists that another reason for poor management and low cuts per hectare is that governments receive too small a share of the returns from logging. On average only between a third and a half of the total resource rent – the difference between the income gained from resource exploitation and the full costs of labour and capital (Mather, 1986) – has gone to governments in most major tropical hardwood producing countries since the 1970s, and the argument is that this encourages short-term exploitation and just a small cut per hectare in order to maximize profits (Repetto and Gillis, 1988). But though government forestry departments should improve their accounting and valuation procedures to gain more income from logging concessions, there is no reason to believe that simply raising the government's share of resource rent would make concessionaires behave more reasonably – indeed the reverse might happen. Nor would logging become less selective, since species exploitation is governed by market demand for tropical hardwood. Finally, there is no guarantee that any additional government income would be ploughed back into reforestation or improving forest management. Most of it might simply go to swell government coffers.

Improve forest protection
Protecting forests is an integral part of good management. Sustainable management is impossible if forests are cleared by encroaching cultivators before they can be logged a second time, or even before they have been logged at all. It is also hindered if timber poachers take out smaller or less valuable trees left behind at the first logging. Concessionaires, whose responsibilities for protecting forests are stated in concession agreements, must clearly place greater emphasis on protection but it is unwise to rely on their efforts alone. Government forestry departments should hire more forest officers to monitor forest incursions on the ground and ensure that offenders are appropriately punished under the law, while at the same time trying to reduce future incursions by supporting local community development (see below). It has been estimated that Indonesia, for example, needs ten times as many foresters in the field as at present, when each of its 17,000 full-time field staff is nominally responsible for looking after almost 7000 ha of forest (Government of Indonesia, 1986; IIED, 1985).

Given the large areas of forest involved, high technology must also be used as much as possible to make the most effective use of the limited number of government personnel, for example, employing remote sensing satellites and light aircraft or helicopters to monitor deforestation and logging. The prospect of greater climatic variability as global climate changes in future due to the greenhouse effect means a higher risk of drought and forest fires in the humid tropics. So monitoring activities should also aim to identify any large fires quickly so they may be brought under control before too much damage is caused.

While establishing formal, legally protected national forest reserves dedicated to timber production is an excellent long-term aim it is impractical in the short term. It was shown in Chapter 3 that only a small fraction of all tropical moist forest is at present within national forest reserves. The demarcation of reserves requires the agreement of other government departments, including those concerned with agriculture and the environment, and this takes time. It also depends on building good relations with local people, for if they do not respect reserve boundaries it is difficult to maintain the integrity of the forest. Forest management should therefore be closely linked with local community development, with foresters actively promoting agroforestry systems and other ways to improve farmers' income.

Improve logging techniques
Much incidental forest damage can be avoided simply by following concession guidelines and good forestry practice but improved techniques would also help. Various initiatives have already been made in this direction. A study in the Malaysian state of Sarawak, for example, showed that damage to the remaining forest during selective logging operations could be halved simply by better advance planning of felling and skidding (Marn and Jonkers, 1981). Careful planning is also at the heart of the Celos Harvesting System, devised in Surinam by researchers from Wageningen Agricultural University in the Netherlands and the University of Surinam. They found that the extra costs incurred by good preparatory work before logging are more than offset later by lower logging costs, not to mention the many benefits that come from halving the damage caused to remaining trees during logging (Jonkers and Hendrison, 1987). Unfortunately, this research has all but stopped because of political unrest in the country and has not yet been implemented on an operational scale (Bodegrom and Graf, 1992). Experience in Brazilian Amazonia, however, shows there are limits to how much logging can actually be improved. After twelve years of trials to identify the most appropriate management practices for the Tapajós National Forest the Brazilian forestry department IBDF (now part of the larger national environmental agency IBAMA) tried to impose tough guidelines on concessions there. The loggers' response was to refuse to apply for these concessions, choosing instead other forests where they were subject to less stringent restrictions (Synnott, 1989).

Centralized or decentralized management?
It has been suggested that tropical rain forest management would be more sustainable if it were transferred from government forestry departments and their appointed concessionaires into the hands of local people. This assumes that local people know more about the forests than anyone else and will be eager to manage for long-term rather than short-term benefits. Forests in the Palcazú district of Peru, for example, are now being managed by a local cooperative, using a system described later in the chapter (Hartshorn et al, 1987) and there are also trials of decentralized management in some Honduran forests (Synnott, 1989).

Although present logging practices undeniably have their faults it is questionable whether more decentralization would be a wise move. There is no guarantee that the logging techniques used by local people will be any better than those of logging contractors today. Management is already highly decentralized, with responsibility for logging given to many concessionaires, and it has proved extremely difficult for forestry departments to monitor and regulate it effectively. The kind of decentralization proposed would mean many more management units and would be even harder to regulate. Furthermore, forestry departments are now being urged, particularly by overseas environmentalists, to improve sustainability according to internationally set criteria. This would appear to require even tighter national control over logging operations and hence more, not less, centralization (Burgess, 1989).

Improve selective logging systems

Selective logging is the almost universal form of logging in tropical rain forests today. It has the advantage of being a relatively low intensity operation, removing only a few trees per hectare, and relies on natural forest regeneration to provide the trees for the next harvest. But there are doubts about present selective logging systems, for though designed with the best intentions they are still based on only a limited knowledge of how forests regenerate after logging. To develop better systems we therefore need to learn more about regeneration through pure and applied research. The low impact on the forest and low cut per hectare of present systems are offset by the need for extensive logging. Would it be preferable instead to increase the cut per hectare to reduce the area of forest logged each year? This question, and the associated issues of marketing and domestic processing, are also discussed here.

Techniques to control deforestation

Undertake more research on forest regeneration

Selective logging has not become the most popular management system for tropical rain forests just because it is inexpensive. Trials of more interventionist silvicultural systems over the last hundred years, described below, mostly proved either unsuccessful or applicable to only limited areas of tropical rain forest, with few advantages over selective logging. But since we know so little about the regeneration of the world's most complex forest ecosystem it is understandable if selective logging systems are flawed. The shortfalls can only be removed gradually as we learn more about how tropical rain forests function.

Since the mid-1970s ecologists have become increasingly interested in studying the regeneration of tropical rain forests after disturbance (Whitmore, 1990). However, most studies have been concerned with regeneration after the creation of natural or artificial gaps in forests and so do not give us all the information we need to improve logging systems, where the degree of disturbance to the forest is different. Longer term studies of regeneration in logged forests are therefore also required for this purpose (Palmer, 1989) and the results obtained should lead to the gradual development of improved selective logging systems.

A number of research projects studying forest regeneration after logging began in Brazilian Amazonia in the 1980s (Siqueira, 1989) and the International Tropical Timber Organization is also supporting projects in other countries (ITTO, 1990a). Governments can help in this research by insisting that certain areas of forest be left unlogged in each concession. These 'strict natural reserves' will provide a valuable baseline against which future regeneration in logged forest can be compared (Roche, 1979).

Increase cut per hectare?

Because of the thousands of tree species in tropical rain forests some selectivity is inevitable in logging operations. But, as Chapter 3 made plain, the small number of marketable species makes the degree of selectivity very high indeed, with cuts per hectare ranging from as low as 5 to (exceptionally) 90 cubic metres compared with a total timber volume of 200 to 300 cubic metres. There have been calls for decades to make greater use of the so-called secondary (or lesser-used) species now left behind in the forest (Booth, 1972, 1978; Plumptre, 1972, 1974). This would raise average cuts per hectare, and though the impact on the forest be more severe and closer to that of deforestation, it would lessen the spread of logging, leading to a reduction in subsequent deforestation by shifting cultivators (see Chapter 2), so the effects would be mixed. It is probably more important to increase cuts per hectare in Africa and Latin America, where they are much lower than in Asia, but too much canopy opening in African and Latin American forests could, ironically, prejudice the regeneration of commercially valuable species (Collins, 1990).

However, the tropical hardwood trade is species-determined, so species are only logged if there is demand for them. Raising the number of commercial species may be attractive in principle, but in practice it is likely to be a slow process, requiring vigorous marketing, particularly by timber trade organizations in tropical countries. Trends in the acceptance of new species are difficult to track due to lack of data but, for example, five species accounted for the bulk of the UK's tropical hardwood imports before the Second World War, a further eight were imported in significant volumes in the 1950s and they were joined by four others in the 1960s and another 12 in the 1970s (Forestry Commission, 1977). But more species logged may not necessarily lead to more efficient forest utilization: the number of species exported from six West and Central African countries in volumes of over 1000 cubic metres rose from 29 in 1951 to 50 in 1973 but the same 30 species that comprised 94 per cent of all exports in 1951 still accounted for 93 per cent of the 1973 total, so the new species made only a minor contribution overall (Erfurth and Rusche, 1976).

Instead of relying on changes in world demand for species tropical countries could help to expand species acceptance themselves by increasing their domestic processing capacity. Many current limitations on species utilization relate to the quality of raw logs, which still account for half of all tropical hardwood exports. If more logs were processed before export local timber traders would have greater influence over which species were logged. Species with potential for use as sawnwood could be tried out on a small scale and if they interested overseas buyers demand for them should grow. Local timber traders could also select inferior species unsuited for high quality surface veneers in plywood and use them as internal veneers instead. Overall, species acceptance and cut per hectare will probably increase gradually, and that will probably be the best outcome for most forests rather than a dramatic rise in intensity, which would lead to even more forest degradation than at present.

Review the role of tropical silviculture

Silviculture is a more interventionist form of management that, instead of relying solely on the natural regeneration of logged forests, manipulates regeneration to promote the growth of commercial timber species in the second and succeeding harvests. It can lead to greater sustainability and commercial productivity. On the other hand, a longer delay is likely between harvests than in selective logging and it causes more degradation of forest ecosystems: some silvicultural systems require removal of the entire canopy, and so cause deforestation.

A number of different systems of tropical silviculture have been tried out over the last hundred years or so in Asia and Africa (Dawkins, 1961; U Kyaw Zan, 1953). Generally, the systems that have been successful have only been so in limited areas, while others might have done well in a physical sense but were so labour intensive that there were not cost effective. However, it is right that their worth be reviewed again in the context of moves towards more sustainable management. The systems are discussed here in three groups: shelterwood systems, the Malayan Uniform System, and enrichment systems.

Shelterwood systems require the forest canopy to be opened and undergrowth, climbers and small trees removed so that seeds of desirable species can germinate under the shelter of selected 'parent' trees. They were developed in Nigeria, Ghana and other West African countries, from the early 1920s onwards, in response to the small numbers of seedlings and saplings in unlogged forests and the unsuitable species composition of regenerating stands (Nwoboshi, 1987). But while the stocks of desirable species did improve growth rates were generally poor and labour costs high. Consequently, these systems were abandoned in Nigeria and Ghana in the mid-1960s (Rietbergen, 1989; Asabere, 1987).

The Malayan Uniform System was developed in the 1950s in Peninsular Malaysia to give a fairly even-aged stand of commercial species in time for a second harvest 70 years after the first (Wyatt-Smith, 1963; Burgess, 1970). Regrowth of seedlings and saplings of desirable species was promoted by heavy felling and extensive canopy opening. Small and medium-sized trees were girdled and poisoned up to six years before the large commercial species were felled. Any remaining undesirable trees were poisoned. Though successful in Peninsular Malaysia's lowland dipterocarp forests it proved unsuitable when tried in the hill dipterocarp forests. However, it is still practised in a modified version in Sabah (Tang, 1987).

Enrichment planting improves the value of the next harvest by planting seedlings of commercial species in either strips or small patches cleared in logged or degraded forests. Trials began in West

Africa in the 1940s but were discontinued by the 1960s because of poor results (FAO, 1958; Asabere, 1987). Experiments were also carried out in the 1950s and 1960s in twelve Latin American countries and extensively in two of these, Puerto Rico and Surinam, but here, as in West Africa, labour costs were high because of the need for weeding (Weaver, 1987).

There is a vigorous debate among foresters about the merits of tropical silviculture (Wyatt-Smith, 1987; Wadsworth, 1987) but the general conclusions we must draw from past experience are that where systems have been successful it has been on a limited scale, and when yields are reasonable costs tend to be high and rarely justified by the returns (Leslie, 1987). The Celos Harvesting System in Surinam referred to above, for example, was the eventual outcome of a research project begun in the 1950s which aimed to apply the Malayan Uniform System to Surinamese forests, but so much human intervention was required that it was too costly to introduce. These and similar experiences elsewhere led tropical foresters to focus on selective logging and plantations from the 1960s onwards.

The poor performance of tropical silviculture in the past was, as in the case of selective logging, influenced by our lack of knowledge of forest dynamics. As we learn more from research so ideas for new techniques will undoubtedly emerge and be tested. Indeed, experiments are continuing in a number of countries. In the mid-1980s, for example, an experiment in strip logging began in the hills of the Palcazú Valley to the east of the Peruvian Andes. In this system all trees in a 20–50 m wide strip of forest are felled on a 30–40 year rotation, each strip at least 200 m from earlier ones so sufficient trees remain to colonize it after felling. Soil erosion is reduced by using animals instead of tractors to remove tree trunks and leaving small branches and leaves on the ground. Management is in the hands of a cooperative composed of local Amuesha Indians, who also own a nearby sawmill. This system is on a smaller scale and more labour-intensive than those elsewhere, so even if it turns out to be successful it may not be widely applicable. Nevertheless, initial results suggest the project is profitable so progress will be watched with interest (Hartshorn et al, 1987).

Expand the area of timber plantations

Over the last 30 years tropical foresters have placed great emphasis on establishing timber plantations (Evans, 1982; Sedjo, 1983). This has been to the general detriment of investment in natural forest management but was justified by the failure of silvicultural experiments and the higher yields per hectare that can be obtained in comparison with selective logging. Compared with average commercial cuts per ha in tropical rain forests of 45 cubic metres in Indonesia and 25 cubic metres in Ivory Coast, for example, a typical high-grade hardwood plantation could yield 225 cubic metres per ha after 35 years and a teak plantation 245 cubic metres per ha after 60 years (Lanly, 1981; Grainger, 1988a). Even higher yields are possible from fast-growing pulpwood plantations, e.g. 150 cubic metres per ha from a plantation of *Gmelina arborea* after ten years, and for tropical pulpwood plantations generally the internal rates of return – a measure of financial profitability (Price, 1989) – are two to four times those in temperate plantations (Sedjo, 1983).

It has been suggested that plantations, because of their high productivity, could eventually supply most tropical hardwood, and that by removing the need for extensive logging of natural forests, with the possibility of subsequent deforestation by farmers, this would help to reduce deforestation rates too. Indeed, according to a rough estimate by this author, compared with the 4 million ha of tropical rain forest being logged each year at the moment, the same amount of wood could be obtained on a sustainable basis from just 28 million ha of plantations.

However, the present plantation area is too small for this to be a viable option in the foreseeable future. Of the 7.7 million ha of timber plantations in the humid tropics in 1980 only 1.4 million ha were dedicated to producing high-grade hardwood similar to that extracted from tropical rain forests (Table 7.3). Over 70 per cent of this area was in Indonesia and a further 20 per cent in Nigeria, Brazil, Thailand, Ivory Coast and Ghana. Another 1 million ha of industrial hardwood plantations consisted of fast-growing trees, such as *Eucalyptus* species and *Gmelina arborea*, grown on short rotations to produce pulpwood and industrial fuelwood and charcoal (Grainger, 1988a).

Although the area of high-grade hardwood plantations grew steadily in the 1960s, the focus of planting then shifted, first to softwood plantations for production of pulp and paper, and then to non-industrial plantations dedicated to environmental protection and production of fuelwood, food and other products for local people (Figure 7.1). The rate at which new forest plantations in the humid tropics were established quadrupled to 0.7 million ha per annum between the late 1960s and the late 1970s, but the share of all industrial wood plantations in the annual planting rate fell from 87 per cent to 58 per cent and that of high-grade hardwood plantations from 33 per cent to just 13 per cent. Currently less than 1 per cent of all tropical hardwood production comes from plantations (Grainger, 1986, 1988a).

Table 7.3. *Areas of forest plantations in the humid tropics in 1980 (thousand ha)*

| | All | Industrial | Industrial hardwood | | |
			All	Sawlogs	Gmelina arborea
All tropics	11,511	7,067	4,349	2,236	138
Humid tropics	7,737	4,580	2,404	1,380	138
Africa	689	407	283	201	67
Asia-Pacific	2,618	1,706	1,174	1,086	0
Latin America	4,430	2,468	947	93	71

Source: Grainger (1986), based on Lanly (1981)

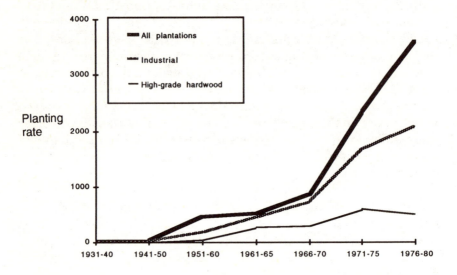

Source: Grainger (1986, 1988a) based on Lanly (1981)

Figure 7.1 *Trends in the rate and type of forest plantation establishment in the humid tropics 1931–80 (thousand hectares)*

The Jari experience

The late Daniel Ludwig launched a great plantation enterprise on his massive 1.6 million ha Jari estate in northern Amazonia in the 1970s. He wanted to produce low-cost pulpwood as feedstock for an on-site pulp mill to take advantage of a projected shortfall in world pulp supplies. Most forestry departments in tropical countries operate under very tight constraints on finance, personnel and equipment. But no technical expertise or expense was spared in establishing the privately owned Jari plantation in the 1970s, and by 1985 it had grown to more than 75,000 ha. So it must count as one of the great pioneering experiments in tropical timber plantations and serve as a signpost to what is possible under ideal conditions.

The initial focus of planting at Jari was *Gmelina arborea*. This is a fast-growing, nitrogen-fixing hardwood tree that can produce high yields of good quality pulp. Unfortunately, it was soon found to be unsuitable for this site and so pines and eucalypts were planted instead, the latter giving the best growth rates (Fearnside, 1988). All the trees have been attacked in varying degrees by disease, however, so growth rates were lower than expected.

Although it does not produce high-grade hardwood, Jari is still a good means of assessing the future of tropical plantations. So far the experience has not been encouraging. The best possible technical support was provided and there were no government subsidies to hide inefficiencies. But high costs, lower than expected growth rates, pest and disease problems and low pulp prices (the projected world supply shortfall never happened) led to heavy financial losses. In 1986, four years after Ludwig had sold the estate to a group of Brazilian companies, the whole Jari estate moved into operating surplus, but this was mainly due to profits from another component – the kaolin mine – rather than the plantations (Fearnside, 1988).

The future contribution of plantations
The Jari experience suggests that we should be cautious about the future contribution of tropical hardwood plantations to the tropical hardwood trade. While there is little evidence of this so far, there are also grounds to fear that wood production might not be sustainable in the long-term if nutrient depletion leads to declining yields over successive rotations (Whitmore, 1990). Another threat is that posed to large monoculture plantations by pests and diseases, something experienced at first hand at Jari. In the background is the ever-present risk of damage by fire: Jari foresters have to keep a close watch to protect their plantations from fires lit by nearby encroaching cultivators.

The share of all tropical hardwood production supplied by plantations is unlikely to rise above 5 per cent over the next 30 years. Indeed, due to the timing of previous planting and delays caused by the long 70-year rotations of the teak plantations established in Asia in the last few decades, overall plantation production could even dip in the early part of next century (Figure 7.2) (Grainger, 1988a).

Because any new high-grade hardwood plantations will take at least 40 years to mature any major initiative to expand their area significantly will not be felt in the next 30–40 years, which will probably be the most crucial period for tropical deforestation. If the bulk of tropical hardwood is going to come from tropical rain forests in the foreseeable future it is essential to reverse the bias against natural forest management and direct more resources to make this more sustainable, along the lines suggested above.

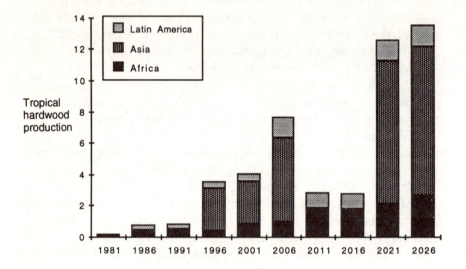

Source: Grainger (1986, 1988a)

Figure 7.2 *Projection of future production of high-grade tropical hardwood plantations 1981–2030 (million cubic metres per annum for five-year periods)*

Plantations still have a major role, so planting should continue, and preferably expand in scale in anticipation of future needs. Projections by this author with the TROPFORM model suggest that tropical hardwood production and exports could peak in the first decade of the next century (see Chapter 9) so more plantations are needed to take on a greater share of all production in the second half of next century. The need is most urgent in Asia, where after the year 2000 timber reserves in the natural forests could become so depleted that exports will decline and supplies will be unable even to meet domestic needs (Grainger, 1986, 1987). As most plantations in that region were established quite recently and are dominated by teak, which has a long maturation period, they will have only a limited ability to cope with this shortfall. Countries in Africa and Latin America have more time to expand their plantations, and the low yields per hectare from forests there should give them a greater incentive to do so.

Minor forest products: conflicts or integration?

The value of latexes, nuts, medicines and other minor forest products gathered by local people was shown in Chapter 6 to be fairly high but continued harvesting is threatened by deforestation and to a lesser extent logging. The recent public focus on the plight of the estimated half a million rubber tappers in Brazilian Amazonia, as they struggled to maintain their lifestyles in the face of expanding cattle ranches, also raised wider policy questions.

Should distinct areas be protected from deforestation and logging and set aside for minor forest product harvesting? A number of 'extractive reserves' have now been established in Brazilian Amazonia, the first in the state of Acre in February 1988 (Fearnside, 1989b). But the human cost was high: the tragic death of the rubber tappers' leader, Chico Mendes, at the hands of cattle ranchers (Byrne, 1990a). Some studies have made grand claims concerning the profitability of such reserves. For example, Mark Peters, from the Institute of Economic Botany in New York, and colleagues estimated that the discounted net present value (see below) of managing a tropical rain forest in Peru for minor forest products was twice that for alternative land uses such as cattle ranches and timber plantations (Peters et al, 1989). But given the poverty of Brazil's rubber tappers, even though domestic prices were heavily subsidised by the government until 1990, such claims should be treated with caution. The Peters study has also been criticized as the forest in question was only 30 km from the Peruvian city of Iquitos and the study used city prices not typical of those in Amazonia as a whole (Barbier, 1991).

Another question is whether logging should be integrated with minor forest product harvesting in the same area of forest. This suggestion has been made from a variety of motives. Some people thought it would make forest management more profitable overall, while others presumably hoped that logging practices would become less degrading. But it is difficult to see how such integration could be achieved in practice. Some studies into this question have been commissioned by the International Tropical Timber Organization but more work is needed. Integration could have a number of disadvantages, e.g. possibly taking responsibility for minor forest product harvesting out of the hands of local people, and causing even more forest damage due to the overexploitation of minor forest products. At present, therefore, it seems that the productivity and sustainability of both practices would benefit if they were kept separate.

Expanding forest conservation areas

This chapter has suggested that the best way to retain a large area of tropical rain forest in the long term is to rely initially on controlling deforestation rather than achieving a rapid increase in forest conservation by setting up a large number of new protected areas. This is for the good reason that such an approach would simply not be feasible. It takes time to establish a protected area well: drawing the boundaries of a new national park on a map is easy, but without adequate funding, personnel and local support the result would be 'paper parks', which look good on maps but whose boundaries are not respected on the ground and that offer little protection to the forests within them. So there is a need to combine traditional 'active' conservation with what we might call 'passive conservation' – action in the agriculture and forest sectors to reduce the rate of deforestation, and good planning to reduce or delay access to forest, e.g. by limiting road building through forested areas. Conservation workers will therefore have the precious time they need to establish new protected areas that have a good chance of lasting over the long term.

Establish priority conservation areas

Another reason for taking time to select future protected areas is that our aim should be to maximize the protection afforded to biodiversity as a whole rather than simply trying to protect the greatest possible area of tropical rain forest of any sort, and since one patch of forest may differ greatly from another in its biodiversity value, selecting in haste could prove regrettable in the long run if valuable species concentrations are left outside the protected area. This has already happened on a number of occasions. It is possible to conserve species outside their natural habitats but given the huge number of species contained in tropical rain forests, many of which remain uncatalogued, conserving whole ecosystems is usually the safest approach. Broad priority areas for conservation have already been identified at international level based on their level of biodiversity and the urgency with which protection is needed. Tropical rain forests on the Pacific coast of Ecuador, for example, have a high priority as they are not only rich in endemic species but also have already been extensively cleared.

Techniques to control deforestation

The main international conservation priority areas can be seen from two ranking exercises. Ten areas identified by the US National Academy of Sciences were the Pacific coast of Ecuador, Brazilian Amazonia, Cameroon, the Tanzanian mountains, Madagascar, Sri Lanka, Borneo, Sulawesi, New Caledonia and Hawaii (Figure 7.3) (NAS, 1980). Another list, by Norman Myers, identified twelve areas that contain a large proportion of all species. It included the Pacific coast of Ecuador and Colombia, the high rainfall foothills of the Andes adjoining western Amazonia, eastern and southern Brazilian Amazonia, the Atlantic Coast of Brazil, Cameroon, the Tanzanian mountains, eastern Madagascar, northern Borneo, Queensland, the Philippines, the eastern Himalayas, Peninsular Malaysia, New Caledonia and Hawaii (Myers, 1988). There is naturally a good deal of overlap between the two lists.

Governments should build on these global rankings to establish national priority areas for conservation, taking into account more specific factors relating to the different kinds of biodiversity described in Chapter 6, i.e. the type of ecosystem, the species content and number of species per unit area, the concentration of endemic species, the genetic diversity of particularly valuable species, and of course the degree of threat.

In some South American countries priority areas have been selected by taking into account the location of proposed Pleistocene Refuges. These are areas to which tropical rain forest is supposed to have retreated during dry periods in the Ice Ages, forming focal areas for speciation and leaving behind today concentrations of endemic species (Figure 7.4) (Haffer, 1969, 1987; Prance, 1973). There is still debate about the merits of the Refuge Theory (Gentry, 1986) and greater uncertainty about the location of possible refuges in Africa than in Latin America (Mayr and O'Hara, 1986). But whether the theory is true or not, it still makes sense to ensure that areas with a high concentration of endemic species are conserved.

Design for long-term viability

There are a variety of types of protected areas. In strict nature reserves, for example, access is limited to scientists and tourists are not permitted. In national parks, on the other hand, the general public is encouraged to visit designated areas (Table 7.4) (Ledec and Goodland, 1988). Whichever type is chosen, when its general location has been decided good planning is needed to ensure it can meet its objectives and be viable in the long term.

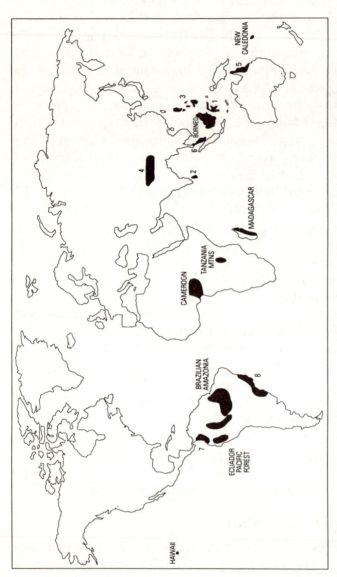

NB. The map is based on the recommendations of NAS (1980b) and Myers (1988). Locations identified by words on the map are those recommended by both sources. Those refered to by numbers only are recommended by just one source: NAS: (1) Sulawesi; (2) Sri Lanka, Myers:(3) the Phillipines; (4) eastern Himalayas; (5) Queensland; (6) Peninsular Malaysia; (7) Colombia's Pacific (Choco) forest; (8) Brazil's Atlantic forest. See the text for further details.

Figure 7.3   Priority areas for conservation in the humid tropics

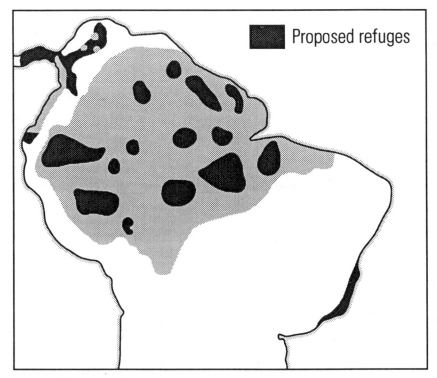

Source: after Prance (1973)

Figure 7.4   Proposed refuge locations in South America

Table 7.4.   Types of protected areas

Type	Characteristics
Strict nature reserves	Entrance for scientific study only, closed to tourism and recreation
National parks	Controlled visitor access
Nature conservation reserves	Limited public access; protection of rare species habitats
Multiple-use management areas	Maintain environmental services, e.g. catchment regulation
Biosphere reserves	Combine core undisturbed zone, buffer zone, manipulative zone and cultural zone
Protected landscapes	Of outstanding natural beauty
World heritage natural sites	Outstanding international importance, strictly protected

Source: based on Ledec and Goodland (1988)

Controlling tropical deforestation

Ecologists have devoted considerable efforts in recent decades to devising methods of determining the optimal sizes and shapes of protected areas that will give ongoing protection to the ecosystems and species they contain. One approach is to estimate the Minimum Viable Population of a given species (Soulé, 1987b). If the population falls below this limiting value, typically of the order of thousands but varying from one species to another, there is a strong chance that the species will become vulnerable to extinction as random events, such as diseases, affect its population. The minimum size of protected area could therefore be that which contains the minimum viable populations of species of particular interest, whether they be plants or animals. On this basis a tropical rain forest reserve intended to protect a minimum of 10,000 individuals of a certain dipterocarp species found at a density of just two trees per hectare would need to be at least 5,000 ha in size, although it would of course protect many other species as well.

Another approach could be to determine the Minimum Critical Reserve Size by applying Island Biogeography Theory and the associated species-area relationship (see Chapter 6). Empirical studies were initiated by the World Wildlife Fund–US in Brazilian Amazonia to monitor changes in species numbers in isolated patches of forest reserves, ranging from 1 to 10,000 ha, that were artificially established like islands in a sea of newly-cleared rangeland on an estate 105 km north of Manaus (Lovejoy et al, 1986). While it has produced some interesting results the project has not lived up to early expectations (Zimmerman and Bierregaard, 1986). But it has shown that the edges of reserves are very susceptible to invasion by exotic sun-loving species that normally live in open fields or the forest canopy. The size of the modified edge region can be quite large: 50 m deep for birds and up to 300 m for butterflies. The significance of the 'edge effect' depends on the size of the reserve. Small reserves are particularly at risk, and even a 50 m edge would modify one fifth of a 100 ha reserve. Conservation planners can minimize the edge effect by establishing large reserves in which the ratio between boundary length and area is kept as low as possible. Circular reserves are clearly ideal in this respect but are not always practical.

The impact of the edge effect can also be reduced by including 'buffer zones' to act as artificial edges. Buffer zones are areas of stable vegetation cover and/or land use, such as managed forest or tree crop plantations, that surround the core protected area and shield it from species invasion and deforestation. In some protected areas the buffer zone is divided into two: an inner buffer zone in which human disturbance is very limited, e.g. to hunting or the collection of minor forest products, and an outer buffer zone in which sustainable agriculture is allowed (Figure 7.5). Buffer zones should be an integral part of conservation activity as they can help to guarantee both the viability of species populations and the support of local people allowed to use them. In the 254,760 ha Cuyabeno Reserve in Ecuadorian Amazonia, for example, which safeguards not only plants and wildlife but also the lands of two Indian tribes, the Secoyo and the Siona, the Indian lands are a valuable buffer between the core of the reserve and the activities of migrant encroaching cultivators (Gradwohl and Greenberg, 1988).

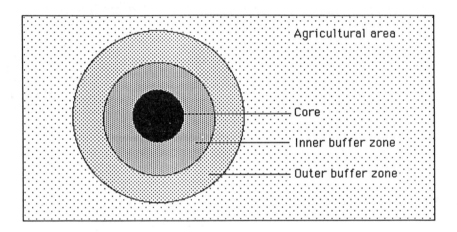

NB The relative sizes of zones do not necessarily match those seen in actual protected areas. The core area is protected. Light use is allowed in the inner buffer zone and sustainable agriculture in the outer buffer zone.

Source: after Ledec and Goodland (1988)

Figure 7.5 *The layout of buffer zones for a typical protected area*

Flexibility in design is also vital for long-term viability. Governments must appreciate that even with the best will in the world and the most efficient planning and protection techniques, some unplanned deforestation is inevitable and a future government might even repudiate some of its predecessors' good intentions. So the sizes of protected areas should be as generous as possible. Another reason is ignorance: the long-term viability of tropical rain forest ecosystems is influenced in ways still imperfectly understood by interactions between different species, including a high degree of plant-animal specificity in the pollination and dispersal of some plant species. This only applies to some plants but there are others, the so-called keystone species (Whitmore, 1990), for example fig trees in Asian and Latin American forests, whose presence is essential to maintaining ecosystem integrity as they are dependable food sources for animals in periods when other species are not fruiting. Finally, conservation planners now appreciate that global climate change could result in shifts in biome boundaries, so that parts of protected areas now covered by tropical rain forest could be overrun by other types of ecosystems. The sizes and boundaries of future reserves should be planned to anticipate this and if necessary allow protected tropical rain forest flexibility to migrate as climate shifts. At present we are unsure how climate will change but hopefully the situation will improve in the 1990s and give conservation planners better indications. In the meantime, where it is possible to establish protected areas that contain a complete range of forest types from lowland to montane, as in Peru's Manu National Park, this would be advantageous.

Involve local people

People living near protected areas can be the greatest threat to their survival, but can also be their strongest supporters. Protected areas are more likely to survive if they have the support of local people, and for this to happen conservation workers should not only consult them before establishing a protected area, but also ensure that the conservation programme of which the protected area is part enhances and stabilizes local activities in farming, tourism, etc. It is particularly important to consult women as well as men, in order to benefit from their practical knowledge, and to involve as wide a cross-section of the local community as possible, so as to prevent possible problems in the future should people feel aggrieved that their opinion has not been taken into account. In this way local people have an interest in seeing that the protected area survives. They may be allowed to continue hunting in the inner buffer zone and practise more intensive uses, such as agroforestry and plantations, in the outer one. The integration of protection and sustainable use is extended even further in the design of the biosphere reserves established as part of the UNESCO Man and Biosphere Programme (UNESCO, 1981; von Droste, 1988).

Approaches to gaining local involvement differ from country to country. The 126,000 ha Korup National Park in Cameroon, for example, is thought to have the highest species diversity of any tropical rain forest in Africa and to cover part of one of Africa's proposed Ice Age refuges (see above). Situated in the south-west corner of the country along the border with Nigeria it contains over 400 tree species. One of the challenges when establishing it was to resettle the people who lived in the park. Officials also tried to win the cooperation of all local people by making the park just one element of a wider regional development programme, which, among other things, advises local farmers on improved agricultural techniques and encourages trials of agroforestry systems (Gradwohl and Greenberg, 1988).

The Cuyabeno Reserve in Ecuador, mentioned above, was designed to increase the income of local people by improving wildlife management and tourist facilities. Giving indigenous peoples full legal rights to their ancestral lands is a good way to ensure that substantial areas of tropical rain forest survive. The Colombian government set an excellent example in this respect when in 1988 it handed over almost half the total area of Colombian Amazonia to fifty tribal groups (Bunyard, 1989).

Gaining local support has been important for the 216,900 ha Khao Yai National Park in north-east Thailand, situated 160 km from the capital, Bangkok, on the south-west edge of the Khorat Plateau. It receives hundreds of thousands of visitors each year. Some villagers living near Khao Yai, which was the country's first national park when founded in 1962, have been given subsidized fertilizers and seeds and other incentives in return for not poaching inside it. The park is also a useful source of income for the villagers who work for it, e.g. as visitor guides, another reason for them to see that it survives (Ewins and Bazeley, 1989).

Paying for conservation

Conservation has to be paid for. Besides the opportunity costs of not exploiting forests for timber, agricultural land or minerals, there are the considerable direct costs of day-to-day management and maintenance, protecting reserves from incursion, relocating any existing forest dwellers, providing educational, accommodation and other services to scientific visitors and tourists, and purchasing the land where it is not already owned by the government. If the area of tropical rain forest within protected areas is to rise sharply in the next few decades there must be a corresponding increase in funding, much of which will have to come from developed countries rather than the tropical countries themselves.

Controlling tropical deforestation

A great deal of money for conservation already flows into tropical countries from abroad. Besides aid from governments and international agencies such as the World Bank, people all over the world contribute to non-governmental organizations (NGOs), such as the Worldwide Fund for Nature, which have achieved a tremendous amount in financing the spread of protected areas in the humid tropics and elsewhere. Innovative approaches, like debt-for-nature swaps, gave a new impetus to large-scale conservation in the 1980s but other new funding mechanisms are needed to maintain the momentum during the 1990s and beyond. This will depend on new economic techniques for assessing the true value of tropical rain forests, the agreement of all countries to accept those values and what they entail, and the development of new national and international institutional frameworks to transfer the necessary funds to developing countries and then use them to support effective conservation and sustainable environmental management in general.

Continue debt-for-nature swaps
Debt-for-nature swaps were a major new feature of international conservation in the 1980s. In return for the government of a tropical country paying for the establishment of a new protected area or the expansion of an old one in local currency, an overseas organization (usually an NGO) purchases some of its foreign debt from banks, usually at a discount and with financial support from other organizations. The country's foreign debt is reduced while the proportion of its territory covered by protected areas is increased – without any loss of sovereignty since the conserved land remains under local ownership.

The first country to benefit from a debt-for-nature swap was Bolivia in 1987. The American NGO, Conservation International, bought $650,000 of government debt at an 85 per cent discount, on condition that the government guaranteed legal protection for the existing 135,000 ha Beni Biosphere Reserve in Amazonia and increased the area protected in that region by creating the 130,000 ha Yacumi regional park, a 225,000 ha watershed protection zone, and a 670,000 ha forest reserve for sustainable timber production. Management costs were paid by the income from an endowment equivalent to $250,000 in local currency. Of this the government only had to contribute $100,000 – what it had cost Conservation International to buy the debt in the first place. The remainder came from the US Agency for International Development. Implementation of the agreement, however, encountered many obstacles, mainly due to public concern about loss of sovereignty (Page, 1989).

Techniques to control deforestation

Since this pathfinding work in Bolivia, other swaps have been arranged in Costa Rica ($5.4 million), Ecuador ($1 million and $5.4 million), and the Philippines ($390,000), usually organized by NGOs but sometimes with the involvement of overseas governments too (Dunne, 1989a, 1989b; Page, 1989). More swaps will be made in future, but the relevance of this approach will probably be limited by how long the Third World debt crisis lasts.

Expand ecotourism
Despite worldwide concern about tropical rain forests and increasing public exposure to their ecological wealth in the developed countries by television programmes, the number of overseas tourists visiting these forests is still quite small. This is understandable given their remoteness, problems with accommodation and acclimatization, and other difficulties. But game parks have been a major source of foreign currency for African countries for years as tourists have flocked to see their wonderful wildlife and great vistas, and the tropical rain forests also have great potential. Tourists pay entrance fees when they visit national parks, and other income comes from spending in hotels and on other local services. Expanding ecotourism could provide some of the money needed to finance further conservation. By enriching local economies in these remote areas, it would also strengthen the support of local people for protected areas. Many tourists could come from inside the country too. For example, besides the 25,000 foreign visitors to Indonesia's national parks and forest reserves in 1985–6 there were two and a quarter million Indonesian visitors (Government of Indonesia, 1986). This not only generates income but also broadens national support for conservation.

In a number of countries, such as Thailand, governments are upgrading the services available to visitors at their national parks, e.g. by improving accommodation and refreshment facilities and supplying better information to visitors on what they can see in the park. The government of Costa Rica gave a new impetus to ecotourism in its country when in 1989 it launched a programme to allow private operators to organize tours inside its national parks (Coone, 1989). We should probably not expect ecotourism to be as successful here as it is in the African savannas, because of the general poor visibility of tropical rain forest wildlife, which perhaps inevitably is more popular with visitors than trees, but it will have a key role in a more comprehensive approach to fund raising.

Assess the full value of forests
Protecting a significant proportion of remaining tropical rain forest

will cost so much money that a fundamental change in conservation financing is called for. New funding mechanisms should be introduced that are based on a more comprehensive economic valuation of all the benefits of conservation. While the most powerful reason for conserving tropical rain forests is undoubtedly still the ethical one, moral persuasion can only expand protected areas so far. Stronger means of justification are needed if we are not to be left with only a small area of tropical rain forest in a pristine state. Instead of simply expressing qualitatively the various benefits of forests, e.g. in stabilizing water flows and climate and housing gene banks, we should try to convert these, however inadequately, into monetary terms, using valuation techniques developed by environmental economists over the last two decades. This could lead to the beneficiaries of conservation contributing to the costs involved, and enable conservation to compete for land with alternative uses on the basis of common economic criteria, furthering the integrated approach to environmental management advocated in this chapter.

Economists have recognized for some time that many aspects of the natural environment are not valued by the market system. Thus, because timber is sold but the role of forests in stabilizing water flows is not, the latter does not have a market value and this can lead to the environmental services supplied by a forest being viewed as inferior to its marketable benefits. When a choice has to be made between retaining forest cover and removing it to plant crops, only a fraction of the full value of the forest may be included in cost-benefit analysis exercises and clearance would seem advantageous. But if the value of the environmental services lost by deforestation is included as a cost in this exercise the result might be very different.

To cope with this 'valuation gap' environmental economists have devised the concept of the *total economic value* of a resource (Pearce and Turner, 1990). This has two parts: *use value* and *existence value*. The use value of a forest, for example, refers to all uses both now (*actual use value*) and in the future (*option value*). These include the sustainable harvesting of marketable goods such as timber and minor forest products (some of which are sold), and non-marketable uses such as recreation and gene banks. These *direct* use values contrast with *indirect* use values associated with environmental services such as catchment protection and carbon storage. If forest were cleared and the land converted to agriculture the present uses would be lost, and we and our descendants would also give up the option of benefiting from them in the future – hence the separate *option value* associated with each actual use value (Weisbrod, 1964). Finally, besides providing goods and services forests are valuable to us just for 'being there'. *Existence value* is what people are willing to pay to ensure that others, both now and in the future, can experience beauty, solitude, and other non-use forest benefits (Krutilla, 1967).

Valuing forests in this way will be a positive, not a negative step for conservation, for it should prevent a lot of forest exploitation happening by default just because it produces some saleable commodities, in contrast to the forest's non-saleable environmental services. John Dixon and Paul Sherman, of the East-West Centre in Honolulu, have made the point that since most protected areas are established in response to the needs of society, if we fail to state the full social benefits from conservation we shall perpetuate its underfunding and, even worse, make it continue to appear like a drain on the public purse (Dixon and Sherman, 1990).

Estimating values
How can we estimate the total economic value of tropical rain forest if only some of its direct uses – harvesting timber and some minor forest products – have market prices? Various techniques have been developed for valuing other direct uses (Dixon and Sherman, 1990). For example, forest economists can estimate the value of forests for recreation using the travel costs of visitors (Price, 1989). Estimating the direct use value of biodiversity is more difficult, however. The commercial benefits from exploiting the gene bank of a single crop can be huge, running into millions of US dollars a year for each new crop variety (Swanson, 1991), but the large number of plants in tropical rain forests and their complex distribution lead to great uncertainty in valuation. Only a small fraction of all the plants have yet been catalogued, the proportion domesticated is even smaller and many potential crops are still unidentified. How can we know if a particular plant will one day be a life-saving drug with a multi-million dollar sales potential? Another valuation problem arises because the impact of deforestation can be irreversible, causing some species to become extinct.

Indirect use values can be estimated in various ways too. One of these employs the market prices of goods and services affected by environmental degradation, for example, estimating the value of crop production lost due to sedimentation of irrigation channels. Another estimates the cost of compensating for impaired environmental services (e.g. regulated water flows) by either artificial means or rehabilitation (e.g. reforesting a catchment) (Dixon and Sherman, 1990). The existence value of forests can be estimated by conducting surveys. If, for example, each person questioned was willing to spend $8 on average to preserve the tropical rain forests of

Amazonia, then assuming a total of 400 million adults in the developed countries the existence value of these forests would be $3.2 billion per annum (Pearce, 1991b).

Realize forest values

Knowing the total economic value of an area of tropical rain forest helps to justify setting up a protected area, but it would be nice if we could realize some of this value to pay for conservation. One difficulty is that most of the people who will benefit from conservation live outside the immediate area and are under no obligation to pay for something they have not asked for. Establishing a protected area of forest, for example, will assure stable water flows for hydroelectric schemes and farmers downstream. On the other hand, if the forest were cleared the same people would have to bear costs due to variable water flows and sedimentation. These are examples of what economists call *external benefits* and *external costs*, respectively.

Within countries it is possible under some circumstances to collect payments from beneficiaries. For example, new accounting systems could be devised so that government agencies and other large organizations contribute to the cost of maintaining environmental services on which they depend and which are paid for initially by another government agency. Thus, conservation of catchment forests could be included in the budget of government hydroelectric or irrigation programmes that use water draining from the catchments. This approach allows external values to be 'internalized' in programme budgets. It does not matter that they cannot be bought and sold on the open market: if programme planners decide that environmental services are essential for the future of the programme, and the government and/or funding agency agree, then the costs can be written into the budget. The World Bank showed how this could be done when, as part of a major irrigation scheme it funded on the island of Sulawesi in Indonesia, it insisted that the Indonesian government establish a catchment conservation area, the 3,000 sq km Dumoga-Bone National Park, to protect water supplies to the scheme (Ledec and Goodland, 1988).

The problems of collecting contributions become greater when the beneficiaries live in other countries. Protecting catchment forests in India, for example, benefits millions of people in adjoining Bangladesh and Pakistan as well. International agreements are therefore needed to compensate countries for conserving forests whose external benefits transcend their own borders. The original reason why the World Bank established its Global Environmental

Facility was to do just that (Riddell, 1990), but its scope has subsequently been broadened and it will now also channel funds to implement key agreements made at UNCED (Pearce, 1992b).

Some of the direct and indirect uses of tropical rain forests, such as biodiversity and global climate stabilization, benefit the whole world, and this makes matters even more complicated. The valuation of biodiversity in general involves immense problems, as already mentioned. These are eased slightly by focusing on particular species that are being commercially developed. The potential economic value of tropical rain forest gene banks is immense and the idea of governments charging fees to plant breeders or pharmaceutical companies that exploit them to produce highly profitable new crop varieties or drugs was mooted as long ago as 1980 in the IUCN World Conservation Strategy (IUCN, 1980). Little progress has been made in putting this idea into practice so far, however. This has been due to difficulties in valuation, uncertainty about the rights of governments over the contents of these forests and the rights of plant breeders or pharmaceutical companies over the products they create from the basic biological material, and fee collection problems.

However, some governments have already entered into agreements with individual overseas companies. For example Merck, the international pharmaceutical corporation, has been given exclusive rights to screen plants, insects and microorganisms submitted to it by Costa Rica's National Biodiversity Institute, INBIO, which will share in the profits made from any subsequent products (Anon, 1992). But before an enforceable worldwide scheme can get off the ground there needs to be international agreement on the respective legal rights of governments and companies to genetic material and derived products, and an international institutional mechanism to coordinate the agreement and channel fees from users to gene bank owners. Though a start was made with the Biodiversity Convention signed at the UN Conference on Environment and Development (UNCED) in 1992, its financial mechanisms are rather general, referring merely to the need to decide fees by mutual agreement and committing signatories to contribute to a new biodiversity fund to finance conservation generally (UNEP, 1992). This funding arrangement, and concern about the implications for intellectual property rights, led the US government to refuse to sign the convention (Pearce, 1992a).

Governments and other bodies that conserve tropical rain forest could be paid for the contribution this makes to mitigating the greenhouse effect. Retaining these carbon stores conveys benefits on

215

everybody in the world, as does protecting regenerating forest and establishing new plantations, which accumulate biomass and serve as carbon sinks. Estimating the value of these roles became easier after studies initiated by the Intergovernmental Panel for Climate Change, which allowed the costs of reducing carbon dioxide emissions by cutting fossil fuel combustion to be directly compared with those of complementary strategies like these. Tropical countries could be paid 'carbon credits' in future to subsidize forest conservation and afforestation (Barrett, 1991) as part of an international effort to stabilize greenhouse gas emissions and, later, their atmospheric concentrations. But this will have to wait until all nations agree on a comprehensive and integrated timetable for action. The developed country signatories to the Climate Change Convention signed at UNCED did not want to make any specific commitments and until they do so the developing countries are unlikely to agree to control deforestation just so the richer nations in the world can carry on consuming fossil fuels as before (UN, 1992; Pearce, 1992c).

Expand the scope of cost-benefit analysis in development projects
The concept of total economic value can also be used by government planners to ensure that projects that clear forest for essential development purposes should pay more heed to environmental impacts than at present. Since environmental components without a market value are often excluded from the cost-benefit analysis used in project appraisal, development projects often appear attractive just because the benefits of the new land use exceed the costs of forest clearance, despite the heavy social costs resulting from lost forest environmental services, recreational functions, etc. (Pearce and Nash, 1981; Barbier, 1991).

To correct for this it has been proposed that the criterion for evaluating projects should be that benefits exceed the sum of costs and the net environmental benefit if the project is *not* implemented (Pearce and Turner, 1990). For projects causing deforestation the net environmental benefit usually corresponds to the total economic value of the uncleared forest. The higher it is, the greater the environmental cost if the forest is cleared and, other things being equal, the less desirable the project. This approach will allow greater discrimination when selecting sites to clear for agricultural and hydroelectric projects but needs the backing of governments and development banks and agencies if it is to be widely used.

Another problem in project evaluation concerns the use of discounting to take account of time. As project costs and benefits associated are normally spread over time, cost-benefit analysis employs the method of discounting to adjust all transactions to the present so they are comparable (Price, 1989). This is because an income of £10 received in ten years time from an investment made today is not the same as £10 received next year. Anyone offered a choice between the two is likely to opt for £10 next year. This is logical, since people generally desire income sooner rather than later, and given an interest rate of, say, 10 per cent per annum the former could be obtained by investing under £4 today but the latter would need about £9. Reversing the compound interest calculation, with the interest rate now the discount rate, £10 received in ten years time is, in 'present value' terms, worth less than £4 today – less than half the present value equivalent of £10 next year.

However, the higher the discount rate the greater the preference given to short-term income. Long term projects, such as those involving forestry, which often last 30, 40 or 50 years, therefore compare badly with those generating a lot of income in a short period by extracting natural resources rather than managing them to ensure future yields. Economists have had a long-running debate over whether discount rates should be lowered to compensate for this and if so how large the reduction should be. (Although, since high discount rates tend to reduce the spread of development generally, low discount rates could mean more development, more rapid resource depletion, and greater environmental impacts (Fisher, 1981)).

According to David Pearce of University College, London, and Kerry Turner of the University of East Anglia, the solution to this problem is not to reduce the discount rate but to add a 'sustainability criterion' to the decision-making process. Projects could therefore go ahead if benefits exceeded costs (valued as above), but only if there were no net change in the total stock of 'environmental capital', as measured by the total economic value of forests and other natural resources (Pearce and Turner, 1990). This approach can obviously lead to difficulties when exploiting an exhaustible resource whose stock by definition cannot then stay constant, or when some environmental degradation is inevitable. To allow for this a project could include an extra compensating component that restores or rehabilitates the environment, e.g. by reforestation, though not necessarily in the same place. If that is impractical, the total stock of environmental capital in a select group of projects could remain constant, and if even this is not feasible, the funding agency could implement additional 'shadow projects' whose sole aim was environmental improvement.

Cost-benefit analysis can also be used before setting up protected areas. One was commissioned for the Korup National Park, for example, by the Cameroon Government and the Worldwide Fund for Nature. With a discount rate of 8 per cent and constant 1989 US dollars the discounted present value of total costs was $15.2 million, of which $11.9 million were direct costs and the rest opportunity costs. Total benefits were $16.2 million, including almost $12 million in direct benefits due to catchment protection ($3.8 million), sustained forest use for local needs ($3.6 million), flood control ($1.6 million), tourism ($1.4 million), and genetic value ($0.48 million). The rest were 'induced benefits' to agriculture and forestry resulting from the integral development project in the buffer zone, described earlier. The net benefit was $1.1 million and the ratio between benefits and costs was 1.07. If the $7.2 million of overseas aid and the loss of $0.78 million in uncaptured genetic value and catchment benefits were included the net benefit rose to $7.5 million and the benefit-cost ratio to 1.94 (Ruitenbeek, 1989).

New valuation techniques like these could do much to control deforestation by providing a more objective basis for generating funds for conservation – e.g. through carbon credits, biodiversity fees and charges for environmental services – and enabling an integrated approach to deforestation control by improving the way deforestation necessary for national development is planned and carried out. The difficulties involved should not be underestimated, however, given the poor quality of even rudimentary land use planning in most tropical countries today, the need for international agreements and institutions to assess the level of international compensatory payments and ensure efficient financial transfers, and the unbelievably partial approaches to controlling deforestation taken so far. As is often the case the mere availability of techniques does not mean they will be used. Whether they are or not depends on the vision and will of policy makers in governments and international agencies.

Saving the forests

Many concerned people and organizations have tried for the last two decades and more to save the tropical rain forests. That they have not been all that successful so far should not disguise the sincerity of their efforts. The task is not easy and it is certainly not the intention of this book to claim that new, quick solutions have now been discovered. What this chapter has argued is that there is a much better chance of success in controlling deforestation over the long term if an integrated approach is adopted that combines action in the agriculture, forestry and conservation sectors. Future programmes should recognize the rights of people in the tropics to aspire to the same process of development experienced by people in the developed countries, while at the same time safeguarding their precious natural heritage. Effective conservation is not easy and requires careful planning and considerable funding, which will depend on devising new sources of finance. Improving the productivity of both agriculture and forestry should not only reduce the pressures that lead to deforestation and stabilize forest boundaries against incursion, but also give conservation workers the vital time they need to expand the area of tropical rain forest under secure protection, and to do it well.

8

Policies to control deforestation

Better agricultural, forestry and conservation techniques will not by themselves control deforestation, even if they are applied in an integrated way. They must be supported by sound policies on the part of governments and international agencies or they will fail. Since the major underlying causes of deforestation and poor forest management are social and economic factors and government policies, policy changes are needed to influence these too. Contrary to the opinions of some commentators, in the majority of countries government policies are not the dominant cause of deforestation, much of which is currently outside governmental control. But deforestation rates could be reduced if governments changed some of their policies on agriculture, forestry and natural resources, and modified their social and economic policies too, though that is more difficult, as most governments probably do not appreciate the eventual impacts that those policies have on the environment.

Prime responsibility for controlling deforestation and improving forest management must rest with the governments of the tropical developing countries, and so the first part of this chapter is addressed to them. But they need assistance from the governments of developed countries, acting both individually and in concert through international agencies, so recommendations are included for them as well. The aim of this chapter is simply to highlight the key policy changes needed. The list of changes (summarized in Table 8.1) has been kept intentionally short in order to focus on priorities.

A major new, and well-funded initiative on the tropical rain forests is essential, though its aims could be achieved mainly through existing organizations and programmes. It is important to learn from the failure of previous international action plans, some of which are reviewed in Chapter 9. Future global initiatives should renounce the partial approaches of the past and adopt instead a more integrated approach to environmental management and

Table 8.1. *Key policy changes for controlling deforestation*

A. Priorities for governments of developing countries
1. Increase support for small farmers
2. Promote large-scale agriculture more carefully
3. Develop an integrated land use policy
4. Revise and publish forest policies
5. Strengthen forestry departments
6. Revise concession agreements and fees
7. Develop a national conservation strategy
8. Ensure a strong environment ministry
9. Improve monitoring of natural resources
10. Cooperate in devising new conservation funding mechanisms
11. Revise social and economic policies
12. Increase environmental component of national development strategies
13. Improve planning techniques
14. Reform land tenure

B. Priorities for governments of developed countries
1. Support positive developments in tropical countries
2. Avoid negative or discriminatory policies
3. Support new conservation funding mechanisms

C. Priorities for international agencies
1. Improve environmental procedures
2. Extend environmental accounting
3. Ensure consistent policies
4. Establish a continuous global satellite monitoring system for the tropical forests

economic development that embraces all land uses, tackles the causes of deforestation and poor forest management, respects the sovereignty of developing countries and acknowledges their right to develop as they see fit.

Tropical deforestation gives us an insight into what will be one of the major policy challenges of the early 21st century: how to reconcile national needs with the responsibilities of global environmental management. Present political preoccupations pale into insignificance in comparison. For those politicians able to rise above national concerns even ensuring global collective military security today is a tortuous affair. Going beyond this to achieving collective environmental security will be an even more formidable task. Some progress, albeit on a limited scale, has been made in the last decades of this century. The world, now faced with the prospect of impending global climate change, is starting to work toward a new international environmental order in which shared responsibilities are matched not just by fine words but by coordinated joint actions too. Programmes to control deforestation should therefore benefit from action to implement, and later extend, the new international conventions on climate change and biodiversity signed, however grudgingly by some countries, at the UN Conference on Environment and Development (UNCED) in Rio de Janeiro on 3–14 June, 1992. We have known of the need to control deforestation for decades but are only now creating the political and institutional conditions necessary to realize our goals. The policy dimension is indispensable and the onus on policy makers is immense.

Policies for the governments of tropical countries

It is the duty of the governments of tropical countries to manage the tropical rain forests and the lands beneath them. The rest of the world can and should support and advise them in this but it cannot dictate what they do: tropical rain forests may have a global value and be the focus of global concern but they are not a global resource in terms of jurisdiction. Here we suggest some key changes to present government policies on agriculture, forestry and conservation to support the improved techniques outlined in Chapter 7 (and summarized briefly at the start of each section), and on social and economic affairs to tackle the underlying causes of deforestation and poor land use. The suggested changes are framed generally so that each government can adapt and embellish them to suit local conditions and priorities.

Make agriculture more productive and sustainable

The agriculture sector remains at the very core of the economies of most tropical countries, so making it more productive and sustainable will not only help to control deforestation but also benefit national development in general. Higher productivity means more food can be grown on less land, reducing the scale of further deforestation. Intensive permanent cultivation should be concentrated on the most fertile soils, with shifting cultivation and other low intensity farming systems encouraged on poorer land. Greater sustainability means that less additional deforestation will be needed to replace farmland abandoned due to falling yields, soil degradation and weed infestation.

Policies to control deforestation

Provide new support to small farmers
The first priority for governments should be to switch the emphasis of their agricultural policies from large farmers to small farmers and help them to manage their lands better. Small farmers account for the vast majority of those engaged in agriculture but have traditionally been neglected by governments. Shifting cultivation has been considered 'primitive' rather than appropriate, and intensive permanent agriculture 'advanced' – even if it turns out to be unproductive and unsustainable. It is time for a change.

More agricultural extension officers should be hired to advise farmers on how to improve their practices by applying techniques that are already known, such as those described in Chapter 7. Governments should seek the cooperation of non-governmental organizations in this task. But research is also needed to devise more productive, yet sustainable, low-input farming systems for both shifting and permanent agriculture.

Focusing on small farmers can result in entirely new approaches to agricultural development. For example, it has been suggested that the fertile alluvial soils on the seasonal floodplains of Brazilian Amazonia, while only accounting for 2 per cent of the total regional area, could support 1 million small farm (10 ha) households, given proper flood control and appropriate farming techniques, in a way that would be far more sustainable than the kind of large-scale clearance that the Brazilian government has promoted in the past. Given the possibility for up to 5 million settlers, compared with a total population of about 11 million people in Amazonia in the early 1980s, the potential for such an initiative is quite significant, provided the environmental impacts can be minimized (Barbier et al, 1990).

To make it worthwhile for farmers to improve productivity, crop prices should be raised where they are regulated by governments and held down in favour of urban consumers. Land tenure reform will also give farmers greater security. Agricultural credit schemes need to be looked at carefully, for while small farmers do need help with finance, too often in the past these schemes have benefited only the larger farmers. The Indonesian government ended subsidized credit in 1990 for this reason but still required banks to give 20 per cent of their loans to small enterprises and farm cooperatives (Murray Brown, 1990).

Promote permanent agriculture more carefully
Planning the expansion of permanent agriculture is still an important duty of government but must be undertaken more carefully

than before. In theory, planned deforestation has many advantages over the spontaneous kind, and should result in forest being replaced by sustainable land uses. In practice, as described in Chapter 2, there are many instances where planning has been bad and new land uses have been introduced that were inappropriate to the sites concerned. Governments have often been obsessed with promoting large-scale agricultural schemes, such as cattle ranching, that have led to extensive deforestation. These are not wrong in principle but in future there must be more careful evaluation of site suitability and better analysis of the economic costs and benefits, including the full environmental costs.

Planning implies the choice of some land uses in preference to others, but too often in the past where governments have promoted the expansion of particular land uses by incentives they have distorted the normal market controls over land management and encouraged unsustainable practices, causing far more deforestation than needed. Despite the huge costs incurred, land use has not been improved, and indeed wasteful resource depletion has taken place (Reppetto and Gillis, 1988). Existing incentives schemes should therefore be re-evaluated and where land uses have been unsuccessful the schemes should be ended, as the Brazilian government has done with its cattle ranching subsidies.

Formulate an integrated national land use policy
The different departments of government should agree on an integrated land use policy for the whole country. This is essential if farming is to become more sustainable and proper legal protection is to be accorded to forest reserves dedicated to long-term timber production and to national parks and other protected areas (Wyatt-Smith, 1987; Poore and Sayer, 1987). Formulating such a policy depends on creating the right political conditions for interdepartmental cooperation, and undertaking a national land capability classification that can be used to determine which areas are suited to which uses. Collective decisions are then needed to zone lands for agriculture, forestry and conservation. A common system of cost-benefit analysis that takes account of environmental values, as outlined in Chapter 7, should be adopted by all departments, so that future projects involving deforestation, whether for agriculture, road building, mining or power plants, can be properly scrutinized before they are given approval. Obviously it is preferable if an integrated land use policy is formulated when a country is still in the early stages of deforestation and forest exploitation but it is equally

useful for countries at more advanced stages, though here the time scale for action will be somewhat shorter.

Make forest management more sustainable

More sustainable forest management will mean less damage to forest by logging, greater protection against illegal logging and clearance, and a sounder basis for perpetuating forest cover. Improving forest management is a long-term process but governments can take various steps to speed it up. Here again we list only the key changes in forest policy needed to do this.

Strengthen forestry departments

The first priority should be to direct more funds into forestry departments so they have sufficient personnel and technology to monitor forests and logging operations, prevent illegal deforestation by improving protection, and ensure that logging operations take place in accordance with government regulations. Indonesia, for example, has been said to need ten times as many foresters in the field (IIED, 1985). Giving forestry departments a higher profile within governments is essential for they have often been the 'poor relations' of other departments in the past, with all that implies for funding and overall policy making.

Revise concession agreements and fees

A complementary change is to increase the length of concession agreements, e.g. to 40 years, to give concessionaires an incentive to manage forests better. Governments will retain the right to cancel agreements if logging operations do not comply with the conditions specified in those agreements. At the same time, the agreements and related fee systems should both be reviewed to ensure that governments receive a higher proportion of the income from logging. The additional income should be used to improve forest management but this cannot be guaranteed. However, the practicality of raising log prices unilaterally to reflect long-term management costs better, as some have suggested, is doubtful, as is the likelihood that it would achieve desired aims.

Publish forest policies

Governments should publish their real forest policies, rather than making do with a fine-sounding set of ideals that are rarely if ever realized in practice. Publication will increase the professional pride and self-confidence of local foresters, encourage them and local people to participate more in the formulation of future policies, and make overseas governments and international agencies more confident in giving financial and technical assistance.

Promote the expansion of forest industries
Governments should promote the expansion of domestic processing capacity so that added value accrues to the national economy, jobs in forest industries are created at home rather than abroad, there is greater local choice over species logged, and sustainable forest management is encouraged by the links established with local industries. Not all governments will be able to do this on the same scale as Indonesia, but a gradual increase in capacity should be realistic. Government subsidies to the new industries are inevitable in the early years, so the costs of these and the loss of revenue from foregone log exports should be taken into account when framing new policies. Governments should also continue to campaign for a reduction in foreign trade barriers to the import of processed wood.

Support pure and applied forest research
Long-term pure and applied research is needed on forest regeneration and forest management, in particular, if improved systems of selective logging and silviculture are to be devised. In promoting this governments should make the best use of their own scientists while at the same time cooperating with other tropical countries to share experiences, prevent duplication and cut costs. In this and other positive developments they should receive international support under the auspices of the International Tropical Timber Organization (see Chapter 9).

Give strong backing to conservation

The overall aim of conservation policy should be to give tropical rain forests the best chance of surviving in the long term. Governments should have a clear strategy and see conservation not as something separate from other economic activities but closely bound up with them.

Develop a national conservation strategy
If it does not have one already, each tropical country should formulate a national conservation strategy that, among other things, lists its priorities for new protected areas, judged on biological criteria and the degree of threat posed to them, and provides a rational framework for future funding and action within the wider framework of a move to more sustainable development. IUCN has suggested that all countries should aim to conserve at least 15 per cent of their original area of tropical rain forest within protected

areas, with another 30–60 per cent inside national forest reserves for sustainable timber production (Collins, 1990). Governments can learn from the experiences of others that have already formulated strategies based on the guidelines in the IUCN World Conservation Strategy (IUCN, 1980), and use the recommendations in the recently published Global Biodiversity Strategy (see Chapter 9) to take specific account of biodiversity conservation needs (World Resources Institute et al, 1992a, 1992b).

Ensure a strong environment ministry
Environment ministers should be given sufficient powers to implement this strategy, influence national development policy as a whole, and bridge the gaps between government departments that inhibit sound environmental management and successful conservation. Fernando Collor's first environment minister on becoming President of Brazil in the spring of 1990 was Jose Lutzenberger, an outspoken environmentalist. He persuaded the President not to extend Highway BR-364 through Acre to the Peruvian border on the basis that it would lead to further huge de-forestation (Gordon, 1990). But he was sacked two years later for claiming that IBAMA, the government department responsible for natural resource management and supposedly under his control, put the interests of landowners first (Homewood, 1992).

Collect better data on national resources
Policy makers and conservation planners can only do their job properly if they have access to sound, up-to-date information. Governments should therefore improve their capacity to monitor national forest cover regularly and establish national biodiversity databases so they know the full natural wealth of their countries and where it is located. Better national monitoring is complementary to the parallel improvement in global monitoring as in most countries local rather than global data needs will probably determine monitoring priorities over the next decade or so.

Brazil and Thailand are leaders in national forest monitoring, as mentioned in Chapter 5, while Thailand and Costa Rica have made good progress in building national biodiversity databases. With funding from the Worldwide Fund for Nature, Thailand is building a national computerized database at Mahidol University in Bangkok to monitor progress in animal species conservation and has so far recorded the national distribution of 916 bird species and 275 mammal species (Round, 1989). Costa Rica's new National Biodiversity Institute was set up by President Arias in 1989 to build a database of all the country's plant and animal species and already

2.5 million insect specimens have been listed (Webb, 1989; Anon, 1992).

Promote passive conservation

Although active conservation – establishing protected areas – is important much can be done by countries to control deforestation by *passive* conservation, which entails wise planning to reduce the threat to forests. One example of this is to limit the building of roads through pristine forested areas. If there is no alternative, roads should be surrounded on either side by buffer zones of tree crop plantations to reduce the chance of clearance. Another form of passive conservation is to recognize the legal rights of indigenous peoples, such as the Yanomami in Amazonia, to their traditional territories. If left to get on with their own lives and given effective government support they will help to protect the forests that are their homes.

Involve ordinary people

Conservation so often appears to be an elitist activity, directed from above without regard to local social conditions. The kind of approach advocated in the previous paragraph should help to overcome such objections but, as the examples in Chapter 7 showed, involving ordinary people at all levels is important if there is to be effective conservation and environmental management. In addition, the democratic traditions of many developing countries are still in a state of evolution but as democracy grows and the capacity of ordinary people to voice their opinions on a wide range of matters is expanded, we should expect much greater participation in the formulation of environmental policies, and this can only be for the best.

Promote conservation as integral to development

Governments should ensure that conservation is not a marginal activity but at the very heart of development planning. A strong Environment Ministry and an integrated national land use policy will contribute greatly to this. But governments can go further and follow the examples of a number of pioneering countries, such as Cameroon, in devising conservation programmes that, through the integrated agricultural development schemes associated with them, benefit rather than disadvantage the local people whose support is so vital for conservation. In that way national park boundaries should be respected rather than being mere lines on a map. The report of the Brundtland Commission in 1987 led to a new awareness of the need for conservation and development to be complementary rather than exclusive, so countries do not have to exploit their natural resources in a thirst for economic growth (World Commission on Environment and Development, 1987). This topic is discussed in more detail below.

Cooperate in devising new funding mechanisms
Governments will be rightly concerned to attract as much money as possible from abroad to pay for tropical rain forest conservation. Debt-for-nature swaps were a welcome boost in the 1980s and some countries are expanding eco-tourism as well. But new funding mechanisms, such as the carbon credits and biodiversity fee schemes outlined in Chapter 7, are needed and the governments of tropical countries should cooperate in introducing them if the world community extends the conventions on global climate change and biodiversity signed at UNCED. Meanwhile, governments could make individual agreements with foreign companies for genetic resource exploitation, as Costa Rica has already done (see Chapter 7). A crucial point to be resolved is the conflict between local sovereignty and global concerns. The governments of tropical countries feel very strongly that they and they alone have the right to say how tropical rain forests are to be managed or conserved and could well refuse to cooperate if new funding mechanisms compromise this sovereignty. It is up to the developed countries and international agencies to be sensitive to this and ensure that such conflict is avoided.

Modify social and economic development policies

Since land use change is an integral part of national development and the ability to manage forests also evolves as society develops, policy changes are needed in areas other than agriculture, forestry and conservation if deforestation is to be controlled. This may seem self-evident but one of the biggest challenges in environmental management is to persuade governments that what happens to the environment is indeed affected by policies that, at first glance, do not seem to be immediately connected with it. Each country's national land use pattern reflects its particular social, economic and political character but also suffers from limitations imposed by the latter. It is not our task here to advocate radical political changes, merely policy changes that could be achieved under present systems as they evolve.

Review social policies
Governments influence deforestation indirectly – and often unknowingly – through the effects of their policies on population

growth and economic development, so they can help to control deforestation by modifying these policies accordingly. In countries such as the Philippines that have high population densities promoting family planning to lower net population growth rates will also help to control deforestation. But not all countries need this remedy – often overpopulation is just a convenient scapegoat for more fundamental problems such as inequitable land tenure (which is definitely a problem in the Philippines) and poverty in general. As part of their population policies Indonesia and Brazil have resettlement schemes to redistribute people from overcrowded areas. While the policies are sound, often implementation is not, and this leads to far more deforestation than necessary. These schemes should continue but with better planning and clearance methods.

Population growth is intimately connected with economic development, which is essential to reduce poverty, although economic growth rates have to exceed population growth rates if progress is to be made in this. Tropical developing countries should not be asked to sacrifice their economic development so that developed countries can carry on as before, or so that tropical resources can be saved to meet global ethical concerns while tropical people remain in poverty. New initiatives are needed to promote urban employment so there will be sufficient jobs for people who would otherwise become encroaching cultivators, but it is also important to make rural areas attractive for those who want to stay there and manage land sustainably.

In countries such as Brazil and the Philippines where inequitable land tenure is a major problem reforms are vital to reduce the number of landless people. But this is easier said than done. President Aquino of the Philippines failed to realize her ambitions in this area during her term of office and progress has also been slow in Brazil. Land use problems will continue to reflect internal economic inequalities, the neglect of indigenous peoples, the concentration of power in the hands of elite groups, and poorly developed democratic controls. Though overcoming these constraints will take time, governments should appreciate that fighting poverty and giving land titles to indigenous and landless people will improve environmental management as well as contributing to social development.

Review national development strategies
It is important that governments review their national development strategies, most of which were designed with very little awareness of their impact on the environment. Future strategies should focus more on the wise management of natural resources rather than seek-

ing to exploit them as rapidly as possible, as in the past. Hopefully the new international focus on 'sustainable development', in developing and developed countries alike, will promote this.

The whole aim of economic development should be to improve human welfare but often economic growth does not result in as much development as expected, as the benefits of growth are not evenly spread and natural resource exploitation causes environmental degradation, which over the long-term impoverishes people and lowers their quality of life. There has been a lot of talk of the need for 'sustainable development' since it was advocated in the IUCN World Conservation Strategy and the Brundtland Report (IUCN, 1980; World Commission on Environment and Development, 1987) but much less about how to achieve it in practice. Clearly, if a country could develop without clearing large areas of tropical rain forest its development path would be more sustainable than one that led to heavy deforestation and widespread land degradation.

One definition of sustainable development is that which 'leads to non-declining human welfare over time' (Pearce, 1991c). In other words, development is sustainable if it is not achieved at the expense of our descendants. Note that the central focus remains the human condition rather than some ideal world where people get richer while the environment is not modified in the process. A condition commonly cited for sustainable development is that the total stock of capital assets, i.e. the sum of environmental capital (e.g. forest resources) and human capital (e.g. buildings, factories, equipment, etc.), should be no less in future than it is now (Pearce, 1991c). An index could be devised to measure the degree of sustainable development by how far the rise in human capital matches the depletion of environmental capital – this being estimated by the total economic value approach outlined in Chapter 7.

In the space of a few years sustainable development has grown from a concept that merely expressed the hope that development could occur in ways that did not overexploit resources and degrade the environment, to potentially a wholly new pattern of development. Much more work is needed, however, for it to be a practical policy option at national level. As it fundamentally challenges the link between resource exploitation, income generation and the main index of growth, the Gross National Product (or GNP), new systems of national accounting are needed to monitor whether development is sustainable (Daly, 1991). Devising them will take time.

Meanwhile, tropical countries have to cope with the realities of everyday life. Governments cannot discard conventional economic management methods for the sake of an attractive but still poorly defined idea. As major changes in national and international markets and national economic accounting are unlikely in the immediate future the most appropriate strategy for governments at the moment is to ensure that all new development projects in a country are as sustainable as possible (Pearce et al, 1990). They should modify regional planning techniques that presently promote the building of new roads, mines, and hydroelectric schemes without any regard to their environmental impact. Environmental impact assessments should be mandatory for all major development schemes. Cost-benefit analyses should include the non-market values of environmental services, such as the regulation of water flows from forested catchments. A better balance should be attained between urban and rural development, for though vibrant urban centres are important, starving the rural sector of funds is a sure way to poorer agricultural practices, less sustainable land use and deforestation. That is one of the more difficult policy challenges facing governments, for agriculture must be sustained at the same time as diversification into the manufacturing and service sectors takes place.

Strengthen institutions at all levels
However enlightened a policy may be, if it is poorly implemented it can result in more, not less deforestation. Lack of trained personnel and institutional capability lead to resettlement schemes that are inadequately planned and implemented, and logging operations that are not properly monitored to check they conform with government regulations. To tackle these and similar problems governments must ensure there is sufficient institutional strength at all levels to implement their policies and programmes successfully. Good policies cannot be formulated unless different ministries talk to each other and agree on a common strategy. New policies on agriculture, forestry and conservation cannot be implemented and enforced without sufficient trained personnel on the ground, and this requires increased provision of education and training places at home and abroad. But central managerial power has its limits, so more responsibility for resource management at local level should be devolved to local communities, including indigenous tribal peoples, and the cooperation of non-governmental organizations, who have a strong grassroots presence and are trusted by local people, should be sought.

Policies for the governments of developed countries

Recognize national sovereignty

It is encouraging to see that the governments of the developed countries have shown more interest in recent years in doing something to control deforestation and improve forest management in the humid tropics. But the tropical rain forests are not global property, so while foreign governments can offer substantial financial and technical support they must do so in a supportive, not dictatorial way, recognize the sovereignty of tropical countries over their natural resources and avoid any hint of 'eco-imperialism'. Vigorous attempts by the governments of the developed countries to prepare a legally binding global forest convention for signing at UNCED failed because tropical countries felt it would compromise their sovereignty and take responsibility for managing the tropical rain forests out of their hands. Instead, only a non-binding 'statement of forest principles' was agreed and even this was very bland (see Chapter 9) (Pearce, 1992d, 1992e, 1992f). This all caused anguish among environmentalists but the fundamental dilemma between global concerns and national sovereignty, already referred to, cannot be resolved so easily. The new international institutional arrangements needed to make global environmental management a reality must be acceptable to all countries and this will require lengthy negotiations.

Support positive developments

Meanwhile, the best way that the governments of the developed countries can help to solve the tropical rain forest problem is to support positive developments. Their already sizable aid commitments could be increased, as appropriate, through both bilateral arrangements with individual countries and contributions to international organizations and programmes. For the most part, in the opinion of this author, no new organizations are needed to tackle the problem, simply more, and better targeted funds directed through mainly existing channels, such as the International Tropical Timber Organization, and the successor to the Tropical Forestry Action Plan, which many governments have used as a focus for their funding in this field since the mid-1980s (see Chapter 9). Some new institutional arrangements will be needed in future if carbon credits and biodiversity fee schemes get off the ground, but initially it is intended that implementation of the Climate Change and Biodiversity Conventions signed at UNCED will be funded through the World Bank's Global Environmental Facility.

Priority areas include giving technical assistance and funding for agricultural development, forestry and conservation planning on the lines suggested in Chapter 7, making sure, of course, that environmental impacts are minimized. Assistance is also needed for scientific research, particularly to improve farming and forestry practices, and gain more knowledge of forest ecosystems, biodiversity and land use change. To improve local skills educational facilities in developing countries should be expanded and scholarship schemes for postgraduate training in overseas institutions increased. Many projects that promote social and economic development and alleviate poverty and inequality could also help to control deforestation but governments should ensure they do not exacerbate it instead.

That caution applies to development projects generally. Too many projects in the past that were supposed to benefit people often degraded the environment, so governments must be specially vigilant in assessing beforehand the likely environmental and social impacts of projects, and refuse to fund projects with severe impacts, such as deforestation or encroachment on to the lands of indigenous peoples. The US Agency for International Development, for example, is now prohibited by law from funding large hydroelectric dam projects and is obliged to safeguard biodiversity when designing projects (Tangley, 1988). Governments can also influence the policies of the international agencies, such as the World Bank, of which they are members (see below), so that projects funded through those agencies are also more environment friendly.

Avoid negative policies

Another imperative is to avoid negative or discriminatory policies. People in developed countries who are in favour of conservation often tend to ignore their fellow human beings in the tropics when criticizing deforestation, seeing it as an aberration that must be stopped but giving little thought to the social consequences. Most developed countries underwent extensive deforestation themselves some time ago so they should be more careful when criticizing current happenings in the tropics.

During the 1980s, for example, some environmental organizations urged the imposition of bans or tariffs on the import of tropical hardwoods, whether in their entirety or from forests 'not managed in a sustainable way'. They believed this would cut the rate of deforestation whereas it could, if adopted, have had exactly the opposite effect. Halting logging would lower the commercial value of tropical hardwood and this, according to the UK House of Lords Select Committee on the European Communities, 'would remove one incentive for conserving tropical forests and managing them on a sustainable basis' (House of Lords, 1990) and so probably increase deforestation rates. Levying import taxes on tropical hardwood, supposedly to raise funds to improve tropical rain forest management, would raise prices and reduce demand. That would benefit the producers of temperate hardwood and softwood, the sales of which would rise, but do little for tropical countries. It would be far better if governments simply allocated sufficient funds from their general budgets to support the work of the International Tropical Timber Organization (see Chapter 9). Discriminating in favour of wood from sustainably managed forests would, in the present situation in which few natural forests are fully sustainable, promote the harvest of timber from plantations rather than natural forests. But plantations are simply incapable of supplying more than a fraction of all demand for tropical hardwood in the foreseeable future, and many were established on former forested land, and so required deforestation! It is also too early to say if plantations themselves will be sustainable in the long term.

So, despite their excellent motives, attempts by foreign governments (and non-governmental organizations) to intervene in the affairs of tropical countries in a negative way can have exactly the opposite effect to that intended, mainly out of ignorance about the actual situation. They should focus on supporting positive developments instead. The governments of tropical countries have for a long time fought the discriminatory trading practices of the developed countries and are likely to see any new moves of this kind as yet further examples of the developed countries acting in their own interest rather than that of the global environment.

Support new funding mechanisms

Much of the cost of conserving large areas of tropical rain forest will have to be borne by the developed countries. Indeed, if the world in general is to approach a more sustainable development path these countries must, paradoxically, maintain a sufficiently high rate of economic growth to be able to afford to change their resource consumption patterns and restore their polluted environments, while at the same time helping developing countries to avoid making the same mistakes (Pearce, 1991c). The governments and individual citizens of the developed countries already contribute a lot of money for tropical forest purposes through government aid agencies and non-governmental organizations such as the Worldwide Fund for Nature, but more is needed. It looked as though a start had been made when 'the Group of Seven', the seven leading industrialized countries, agreed at their 1990 summit to establish a $1.5 billion fund for tropical rain forest conservation in Brazil. After years of suspicion of outside intervention by previous governments, President Collor gave a warm response to this and welcomed the prospect of future debt-for-nature swaps too (Lamb, 1991; Johnson and Fidler, 1991). However, he became angry when, a year later, only $35 million had been committed to the fund (Lamb, 1992).

Two new funding mechanisms were suggested in Chapter 7. Paying tropical countries to conserve their forests to mitigate the greenhouse effect would complement action by the developed countries to reduce fossil fuel combustion. But the tropical countries will probably not proceed with this scheme unless the developed countries first commit themselves to a timetable to reduce their greenhouse gas emissions – which they failed to do in the Climate Change Convention. Similarly, the Biodiversity Convention did not include specific arrangements whereby companies utilizing forest gene banks to develop new crops or drugs could pay for that privilege. Many of the developed countries that signed the convention did not like its open-ended commitment to a general biodiversity fund or how this would be allocated, and the USA refused to sign (Pearce, 1992a, 1992d, UNEP, 1992). So there is some work to do before the developed and developing countries can agree on the kind of international financial transfers that will make global environmental management a practical proposition.

Policies for international agencies

Governments also channel aid through international agencies such as FAO and the World Bank, which, by virtue of the huge amounts of money they lend or donate for development projects, have an immense influence on environmental management in the humid tropics. In the past, through their funding of massive highway construction projects or hydroelectric schemes, this influence was often detrimental and agencies must try to prevent similar mistakes happening in future.

Improve environmental procedures

It is therefore essential that international agencies improve their environmental procedures. Fortunately, the World Bank, and other agencies in turn, have increasingly recognized their past failings and are revising their policies and programmes accordingly. During the 1980s, the World Bank, for example, published new policies on tribal peoples, wilderness areas and dams and reservoirs, promoted new thinking in environmental economics and set up a new Environment Department to oversee this aspect of its operations (Ledec and Goodland, 1989). It and the other development banks should build on this to ensure that environmental impacts are taken into account at all stages of the project cycle from appraisal (or formulation) to implementation. Non-market values, like those of forest environmental services, which are usually neglected in pre-project cost-benefit analysis, should now be included as suggested in Chapter 7. Development banks should also, as appropriate, internalize external costs in project budgets or arrange that they be compensated by environmental conservation or restoration projects elsewhere. This happened in Indonesia when a new national park was established in Sulawesi at World Bank insistence to protect water flows from a catchment that supplied a major irrigation scheme funded by the bank (Ledec and Goodland, 1989).

Extend the scope of environmental accounting

There are other possibilities for development banks to cost the values of environmental services not priced by markets and, even, for example, to devise a system of environmental credits transferable between countries. This is certainly needed when the environmental services of one country, e.g. river flows from forested catchments, benefit another. The World Bank's new Global Environmental Facility (GEF) was set up to distribute funds for this purpose (Riddell, 1990) and it has been proposed as the initial channel through which funds will pass to implement the Climate Change and Biodiversity Conventions signed at UNCED. It is hoped that these will be extended to include more sophisticated funding mechanisms, like those discussed above, and there is also scope to use the GEF as the basis for a new system of environmental accounting and international financial transfers that could eventually make sustainable development a reality on the international scale.

Adopt consistent policies

United Nations agencies, such as FAO, UNEP and UNESCO (the UN Educational, Scientific and Cultural Organization), provide considerable technical support, training and information sharing in tropical agriculture, forestry and conservation and this work should continue. But it is not uncommon for the policies of different agencies, and even different departments of a single agency, to be in conflict. Thus, the agriculture department of an agency such as FAO can easily promote deforestation at the same time that its forestry department is trying hard to control it. It is vital that UN agencies avoid this and have internally consistent policies.

Establish a continuous monitoring system for the tropical forests

Effective international policy making and implementation is impossible without good information. UN agencies have traditionally performed a vital role in collating essential international data and given the limited abilities of many tropical countries to survey their forest resources frequently this role still remains essential. But though the UN Conference on the Human Environment in Stockholm in 1972 called for a continuous global monitoring system for the world's forests this has still not been established. The result has been a general lack of reliable data on tropical forest areas and deforestation rates, as described in Chapter 5.

This shortfall must be reversed by establishing a continuous global satellite-based monitoring system for the tropical moist forests as soon as possible, preferably with the UN taking a leading role, and extending this later to monitor other forests and other global environmental problems, such as desertification. If the UN does not give a lead and ensure that the FAO 1990 Tropical Forest Resource Assessment Project continues on an expanded basis, under the auspices of FAO, UNEP, or perhaps a new dedicated UN global monitoring organization, it is quite possible that an NGO or scientific group will take on the monitoring role that ideally should be a UN responsibility (Grainger, 1990e). IUCN has already taken some steps in this direction through its mapping work (Collins, 1990).

In the medium term a centralized global monitoring system is probably the most practical option, for as national monitoring programmes develop they will tend to focus on collecting data to meet specific national priorities, rather than attempting frequent comprehensive monitoring of the kind needed by global programmes. There is also still great sensitivity on the part of developing countries to releasing data about their natural resources. But should a government have the right, for example, to withhold recent data on its country's forest area, or should its prerogative only be confined to more sensitive commercial data, such as the volume of commercial timber in forests? International agreement is needed to define the limits of sovereignty over natural resources data needed to monitor global environmental change.

Increase international cooperation on the environment

International environmental policy is at a watershed. Previously the nations of the world have been content to sign agreements at international conferences and then either decide their own environmental policies within the framework provided by these agreements, or choose to ignore them. But if we are to begin to truly manage the global environment in order to consciously control our depredations this will no longer be sufficient. The world community will have to commit itself collectively to meeting global targets, and to do that each nation will have to decide on a legally binding national target. This is such a major advance that most nations are simply not ready to take it yet if it threatens their economies or compromises their sovereignty. The Montreal Protocol on reducing the depletion of the ozone layer by reducing production of chlorofluorocarbons was agreed because the threat was shown by satellite imagery to be imminent and the remedy was possible without too great a cost. But the developed nations would not agree at UNCED to a timetable to reduce their carbon dioxide emissions and the developing nations would not agree to a forest convention because in both cases the costs were too high.

Saving the tropical rain forests requires coordinated national action within an international framework. The forests are not a global commons like the atmosphere, even though what happens to them has global effects, and so the sovereignty issue is more contentious here. International action will have to be achieved initially by means of non-binding agreements that take a step-by-step approach to solving the problem while ensuring increasingly greater cooperation between tropical countries and developed countries. The United Nations is best qualified to suggest at each stage how to reconcile national sovereignty and global needs, as it showed when the International Tropical Timber Agreement, which furthers the common interests of tropical hardwood producing and consuming countries, was successfully negotiated under the auspices of the secretariat of the UN Conference on Trade and Development (UNCTAD). But there is a big gap between tropical countries being ready to accept technical and financial assistance for their development programmes, on the one hand, and agreeing to modify their

239

economic activities for global ends, on the other. Overcoming this obstacle will take time and patient diplomacy on the part of UN officials. Tropical countries are unlikely to accept the kind of restrictions on their activities proposed in the abortive Forest Convention. Any agreement reached must recognize their sovereignty and right to develop as they see fit, and place their progress towards a more sustainable development on a par with that of the developed countries, for both are still on a learning curve. The UN must facilitate this growth process in planetary management but to do so it requires firm support and funding from its member countries.

Policies – the heart of the matter

Deforestation cannot be controlled or forest management improved unless the governments of tropical countries make some fundamental changes in policy. To do this they will have to make a major conceptual leap to avoid thinking of deforestation as merely a 'forestry problem', an error in which they have actually been encouraged by both foresters and environmentalists from developed countries and which is compounded by the traditional compartmentalization of agriculture, forestry and environment ministries. Changes are needed not only in agriculture, forest and conservation policies, but also in social and economic policies, often not associated with deforestation, which influence the major underlying causes of the problem.

The governments of developed countries also need a broader vision. Their policies should focus on supporting positive developments rather than trying to restrict how developing countries run their affairs. Their support for deforestation control programmes should cover not just forestry and conservation but also agricultural development and socio-economic development in general. They should promote the greater involvement of non-governmental organizations as well.

The role of international agencies is also vital. The World Bank, and other development banks, have already made great strides in improving the environmental aspects of project design and implementation, as have the national aid agencies of developed countries. They should carry on along the same lines and help to devise and implement new global conservation funding mechanisms. Working with other UN agencies they can provide a sound framework for international action to control deforestation and improve forest management.

9

Synthesis, progress and prospects

The deforestation of tropical rain forests is one of the most important environmental problems in the world today but, as this book has shown, it is a rather complicated problem and by no means easy to control. The first part of this chapter sums up the book's main findings. This is followed by a review of progress made so far in controlling deforestation at national and international levels. The chapter ends with some thoughts about future trends and prospects.

Synthesis

The complexity of the tropical rain forest problem is what might have been expected from a vegetation type that surpasses in extent the land area of the USA, stretches over three continents and numerous islands on either side of the Equator and is found in more than 60 countries, each varying in economic conditions, culture and historical background. Tropical rain forest itself is certainly not a homogeneous vegetation cover on the global scale, comprising a variety of forest types and differing in species composition from place to place. This provides a challenge to conservation planners, and as human impacts vary between countries it is also difficult to generalize about the causes and consequences of deforestation.

That being said, two main human impacts on tropical rain forest were identified in this book: deforestation and selective logging. Deforestation is caused directly by the replacement of forest by another land use. A wide range of agricultural and other land uses are involved, including different types of shifting and permanent agriculture, mining, hydroelectric schemes, etc. But the changes in land use take place in response to various underlying causes, including population growth, economic development, poverty, government policies and land use unsustainability.

Because selective logging does not clear the forest it does not cause deforestation as defined here, although poor logging practices can

241

result in extensive forest degradation and pose a serious threat to sustainability. However, logging is an indirect cause of deforestation in that farmers might invade concessions after loggers leave and clear the forest for cultivation. So making forest management more sustainable, which requires that forest be properly protected from such illegal clearance, will therefore help to control deforestation as well.

The tropical rain forest problem arises partly from the expansion of human activities in forested areas, and partly from the poor way in which lands and forests are managed. This book has suggested that the expansion of agriculture and logging is an inevitable if regretable consequence of population growth and economic development, just as it was in temperate countries centuries ago, although economic development can also help to control deforestation by generating the money needed to make agriculture more productive. Poor farming practices, on the other hand, occur because of people's poverty; unequal access to land; the general difficulty of farming sustainably on poor tropical soils; the effects of economic development, which can reduce the sustainability of traditional practices; the use of inappropriate new techniques; and misguided government policies that introduce land uses into areas to which they are not suited.

Some blame for current shortcomings in forest management must rest with the management systems employed, but mostly they result from the way in which the systems are implemented, with concessionaires circumventing government regulations and government foresters unable to monitor and regulate their activities properly. At the moment the tropical rain forests are being mined for their timber rather than managed as renewable resources, and while this represents a failure of forest policy, it is nevertheless typical of the early phases of forest exploitation generally – similar patterns were seen in temperate forests in the past. So however concerned people from the industrialized countries may be about what is happening to the tropical rain forests, they are not really in a position to criticize tropical countries for repeating their own experiences, and should try to understand the development process that these countries are going through.

Tropical deforestation is part of a process of land use evolution that goes back a long way. At some point in the history of many tropical countries it was influenced by the colonial rule of temperate countries such as the UK, France and Spain and the effects of this can still be seen today. Sadly we lack reliable global data on historical trends in the area of tropical rain forest. What is totally reprehensible, however, is that current estimates of forest areas and deforestation rates are so inaccurate, even though data collected by satellites and other advanced remote sensing techniques have been available now for 20 years. Until 1992 estimates referred only to tropical moist forest – tropical rain forest and tropical moist deciduous forest combined – rather than to tropical rain forest alone.

Compared with the 1,081 million ha of tropical moist forest that remained in 1980, about 6 million ha was deforested and 4 million ha logged annually in the late 1970s. The deforestation rate is thought to have risen in the 1980s, according to one estimate to as much as 14 million ha per annum, but Chapter 5 suggested that this was probably an overestimate and FAO recently released a lower figure of 12.2 million ha per annum. Charting recent trends in total tropical moist forest area is not a very helpful guide to trends in deforestation, for from between 900 and 1000 million ha in the 1970s it seemed to increase to 1081 million ha in 1980, while estimates for 1990 placed it at 800 million ha on the one hand and (tentatively) at 1282 million ha on the other. If deforestation rates were to continue unchanged these figures suggest that on the most optimistic scenario all tropical moist forest could disappear in 170 years, and on the most pessimistic scenario in just under 60 years.

Such poor estimates result from infrequent national forest monitoring by the governments of tropical countries, combined with the international community's failure to compensate for this by installing an effective global monitoring system. Lack of reliable data hinders the design of policies and programmes to control deforestation and improve forest management, as policy makers do not know how urgent the problem is or where the priority areas for action are. The quantity and quality of data should improve over the next few years, but probably the most we can expect are reliable estimates of forest areas and deforestation rates, perhaps divided into the main forest types. Yet these were needed 10 to 15 years ago! Owing to the urgency of the threats to global climate and biodiversity caused by tropical deforestation, far higher resolution data are now needed on an annual basis covering changes in tropical forest biomass and biodiversity if we are to respond to current concerns about these issues. But there is no sign yet that the kind of continuous global monitoring system that could provide such data, and called for as long ago as 1972 at the Stockholm Environment Conference, will become operational in the foreseeable future.

The deforestation of tropical rain forests is having a number of unfortunate consequences. Indigenous forest-dwelling peoples are being decimated. A large quantity of valuable timber is being burnt before it can be utilized commercially. Deforestation and logging are eroding biodiversity, threatening the genetic diversity of many crop plants, minor forest products, and plant species that could be of use in the future but have not even been catalogued yet, and making between five and 16 species extinct each day. Deforestation and logging are also degrading soil, changing water flows from catchments, and leading to the sedimentation of rivers, reservoirs and irrigation systems. These impacts are considerable and threaten the reliability of water supplies to downstream farmers and other water users, but there are insufficient data to link deforestation with the flooding of large rivers. Deforestation is also contributing to the enhanced greenhouse effect, which is expected to increase average global temperatures and change precipitation patterns. Between 7 per cent and 30 per cent of all carbon dioxide emissions come from tropical deforestation. The effect on the oxygen content of the atmosphere is regarded as minimal, however, and the possible effects on local climate are uncertain.

To control deforestation, this book has advocated an integrated approach combining action in the fields of agriculture, forestry and conservation – something not attempted by previous international programmes. The underlying causes need to be tackled as well as the land uses involved, and this means changes in social, economic and development policies as well as in agriculture, forestry and conservation policies. We cannot stop the evolution of land use in tropical countries, but what we can do is to attempt to reduce its negative aspects.

Action in the agriculture sector should aim to make farming more productive and sustainable. Given the large areas of poor soils in the humid tropics, only limited areas are suited to intensive agriculture so sustainability is a major problem. Lack of sustainability represents a failure for tropical countries because more deforestation is needed to maintain food production and the benefits obtained from the original deforestation are lost. The more fertile lands in each country should be identified so that intensive agriculture can be concentrated there, reserving the remainder for forest management, low intensity shifting cultivation and agroforestry systems.

Selective logging is a logical way to manage tropical rain forests given that only a small proportion of the thousands of tree species are commercially marketable. Owing to the relatively long rotations required, high-grade hardwood plantations cannot quickly take over the bulk of tropical hardwood production, so there seems no alternative to placing greater emphasis on natural forest management if tropical timber supplies over the next 50 to 100 years are to be assured. But this all depends on management practices becoming more sustainable than they are at the moment. Significant advances could be achieved over the next 10 to 20 years by improving the implementation of existing selective logging systems through a combination of better monitoring and regulation, incentives to concessionaires and more government investment. Developing better management systems will take longer because of the need to undertake research and monitor current practices.

While managing tropical rain forests for timber production is an excellent way to maintain large areas of forest free from clearance, logging does disturb the forest so substantial areas should also be conserved, free from all uses, in protected areas. But conservation planning is not easy either. There are a variety of types of tropical rain forest and this vast green carpet covering the humid tropics masks a startling heterogeneity in the distribution of plant and animal species. Our aim should not be just to conserve large areas of tropical rain forest anywhere, but to ensure that those areas of forest chosen for protection contain the richest species concentrations and are sufficiently large that their biodiversity has a good chance of surviving indefinitely. This requires good planning, and careful establishment of protected areas so they are respected by local people and outsiders alike. To give conservation workers time to do this governments should try to alleviate the pressures leading to deforestation, by taking action in the agriculture and forest sectors, tackling the underlying social, economic and political causes of the problem, recognizing the land rights of indigenous peoples, and minimizing road access to forested areas as much as possible.

However good the intentions of governments may be, their ability to control deforestation, improve forest management quickly and ensure the integrity of protected areas is still rather limited. For while they can reduce the deforestation caused by their own policies a lot of other deforestation is outside their control, and it is very difficult for them to protect forests and regulate logging properly on such a huge scale. So a government may have excellent policies but be unable to implement them successfully because of shortfalls in personnel and organizational capability that will take some time to overcome. Such delays can be reduced by strengthening government institutions but they cannot be avoided altogether.

Progress in controlling deforestation

Progress in agriculture
What do we have to show for more than 20 years of concern about

tropical deforestation? In the agriculture sector, relatively little, which is not surprising as its role in causing the problem has been neglected. Three important advances are the development of high-yielding varieties of rice and other crops, the rapid growth in research into agroforestry systems since the 1970s, and the start of research into low-cost modifications of shifting cultivation. The end to cattle ranching subsidies in Brazilian Amazonia is also an advance of sorts. High-yielding varieties have their good and bad points, for while they enable more intensive production on the best lands they have not been adopted widely in all countries since small farmers cannot afford the high cost of seeds, fertilizers and pesticides, which have also had environmental impacts. Agroforestry is not a universal remedy for deforestation either, but it does provide an important basis to improve the productivity and sustainability of agriculture in ways that are compatible with local environmental conditions and do not require outright replacement of existing farming practices.

Progress in forestry

In the forest sector, progress has also not been impressive, but the issue of sustainable management has been raised and adopted as a major world policy goal. The International Tropical Timber Organization has initiated some useful research programmes on key topics, many of which still have to bear fruit. It would be wrong to think of the rate of reforestation as a prime measure of progress in controlling deforestation. For most reforestation consists of the establishment of plantations to satisfy demand for timber, not to compensate for forests lost to agriculture, so comparisons between the rates of deforestation and reforestation do not have much meaning. However, progress in plantation establishment has been slow.

Progress in conservation

In the conservation sector there has been a slow but steady increase in the amount of tropical rain forest within protected areas, made possible by the improved awareness of governments, public pressure and the use of innovative funding mechanisms such as the debt-for-nature swaps described in Chapter 7. The general orientation of conservation has gradually changed over the last 20 years from focusing on the conservation of threatened animal species to building networks of protected areas containing examples of all the major ecosystem types in each region to planning conservation in such a way as to conserve the maximum amount of biodiversity on the global scale, an approach not incompatible with the first two.

Future conservation activities will benefit from long-term ecological research into how large protected areas have to be to ensure sustainable conservation, and also from work by environmental economists to devise new valuation methods.

It is difficult to give an exact figure for the area of tropical rain forest that has been conserved already but it does not exceed 8 per cent of the total. Data from the International Union for the Conservation of Nature (IUCN)[1] show that 42 leading countries in the humid tropics had protected areas covering 87.3 million ha in the late 1980s (Table 9.1) (World Resources Institute, 1988). This was equivalent to 8 per cent of the total tropical moist forest area but of course not all of it was forested. Another estimate from IUCN staff is that about a third of the 2000 protected areas that cover more than 200 million ha in the tropics contain some tropical rain forest or tropical moist deciduous forest but it is impossible to say more than that because of lack of data (Collins, 1990). By 1988 23 biosphere reserves had been established in the humid tropics under the UNESCO Man and Biosphere Programme. They covered a total of 5.8 million ha but, by the very nature of biosphere reserves, only part of this area is strictly protected, the remainder being devoted to sustainable development (Oldfield, 1988).

Table 9.1   *Protected areas in the humid tropics by region (million ha)*

Africa	26.4
Asia-Pacific	20.8
Latin America	40.1
All humid tropics	**87.3**

The countries that have achieved most in conservation may be identified by various rankings. Sri Lanka, Panama, Senegal, Ecuador, Costa Rica, Indonesia, Thailand and Venezuela all have over 7 per cent of their territories protected. Indonesia, Thailand and Brazil each have more than 50 protected areas. Indonesia, Brazil and Zaire, which contain half of all tropical moist forest, account for two fifths of the total protected area in the humid tropics (Table 9.2). It is notable that Indonesia and Thailand are among the leaders in all three rankings.

Countries that have achieved least in conservation include Burma, Equatorial Guinea, Guinea Bissau, Kampuchea and Laos (no protected areas); and Guyana, Guatemala, Guinea, Nicaragua,

Table 9.2. Progress in conservation in the humid tropics

National protected areas and percentage of total protected area in the humid tropics

	Area (ha)	Percentage of total
Indonesia	13,590,792	16
Brazil	11,929,634	14
Zaire	8,827,000	10
Venezuela	7,388,912	8
Peru	5,354,328	6
Colombia	4,932,765	6
Bolivia	4,837,143	6
Thailand	4,015,912	5
Central African Rep	3,904,000	4
Ecuador	2,627,400	3

Percentage of national land area within protected areas

Senegal	11
Panama	11
Sri Lanka	11
Costa Rica	9
Ecuador	9
Thailand	8
Indonesia	8
Venezuela	8
Gabon	7
Uganda	7

Numbers of protected areas in countries

Indonesia	135
Thailand	65
Brazil	51
Malaysia	38
Sri Lanka	37
Venezuela	34
Colombia	32
Madagascar	31
Philippines	29
Costa Rica	21

Sources: World Resources Institute (1988), Ledec and Goodland (1988)

Papua New Guinea and Vietnam (less than 1 per cent of their territories protected). Countries in which protected areas cover only 1 per cent of national land area and where more progress seems needed include Brazil, El Salvador, Liberia, Nicaragua, Nigeria and Sierra Leone.

Brazil, with 30 per cent of all tropical moist forest and by far the largest country in the humid tropics (which helps to explain its low percentage rating) has a unique conservation responsibility. Its protected areas already cover 14 million ha overall, with 11 million ha in Amazonia, where there are five national parks, seven biological reserves and six ecological stations. They cover only 2 per cent of Brazilian Amazonia but the number of protected areas rose significantly in the 1980s (Padua, 1989). Costa Rica has 9 per cent of its territory protected but it is a smaller country and has been extensively deforested: forest cover was down to just 17 per cent by 1983 (Sader and Joyce, 1988; Sun, 1988). Zaire has an estimated 10 per cent of all tropical moist forest but little is known about the extent of its forests. As the former Belgian Congo it established the first national park in Africa in 1925 – the former Albert National Park – and now has four national parks covering 5.6 million ha (Kendrick, 1989; Pullan, 1988).

The achievements of previous international initiatives

Tropical countries need substantial international financial and technical support from developed countries if they are to control deforestation. There have been a number of concerted international initiatives to tackle this problem, but they have not been altogether successful. Five action programmes are reviewed here: the World Ecological Areas Programme, the World Conservation Strategy, the Tropical Forestry Action Plan, the International Tropical Timber Agreement and the Global Biodiversity Strategy. The relevant achievements of the UN Conference on Environment and Development (UNCED) are also assessed.

The World Ecological Areas Programme
The World Ecological Areas Programme (WEAP) was published in a special issue of the *Ecologist* magazine in 1980 with the backing of prominent environmentalists from all over the world (Goldsmith, 1980). It was very much ahead of its time, and anticipated in some respects the recent debt-for-nature swaps mentioned in Chapter 7. The basic principle was that governments that conserved large areas of tropical rain forest would be given loans to pay for this and the establishment of timber plantations. The latter would eventually produce timber more efficiently than natural forests and help to pay back the loan, the collateral for which was the area of tropical rain forest to be conserved.

WEAP gave tropical rain forest an intrinsic monetary value for the first time and proposed a way of financing conservation that was innovative and made commercial sense. One drawback was what might happen if the loan had to be called in. Could an outside organization take over large areas of a country's territory? If so, what political problems might that cause? WEAP was never implemented. It was relaunched in revised form in 1987, with countries now being relieved of their debts in return for conserving forest, but it drew criticism from many quarters and so again did not make any headway (Goldsmith et al, 1987).

The World Conservation Strategy
The World Conservation Strategy (WCS), launched by IUCN in 1980, aimed to provide a framework for conservation by governments and international agencies (IUCN, 1980). Each country was urged to devise its own national conservation strategy. The underlying philosophy of the WCS was the new principle of sustainable development, outlined in Chapter 8 and essentially an approach to development that takes account of the need for conservation and vice versa. National action was central to achieving the strategy's international priorities, which included conserving tropical moist forests and genetic resources in general. It was suggested that governments and corporations that derived income from exploiting genetic resources should pay for forest conservation.

The WCS also came under criticism. Some commentators were unimpressed by the major shift towards recognizing the validity of development and saw it as 'paying lip service to the social context' (Anderson and Grove, 1987). Since it was only a framework for action it is difficult to measure its success in terms of actual accomplishments, but 35 countries now have national conservation strategies (World Resources Institute, 1988). In 1991 IUCN published an updated strategy, called 'Caring for the Earth', which proposed a multi-billion dollar action plan for consideration at the UN Conference on Environment and Development (UNCED) (Hunt, 1991; IUCN, 1991).

The Tropical Forestry Action Plan
During the second half of the 1980s world efforts to conserve tropical forests generally focused on supporting the Tropical Forestry Action Plan (TFAP). This had two separate origins. The first was a plan published in 1985 by the World Resources Institute, the World Bank and the UN Development Programme, called 'Tropical Forests: A Call for Action' (World Resources Institute et al, 1985). The second, actually called 'The Tropical Forestry Action Plan', was published in

the same year by FAO (FAO, 1985). In 1986 the two plans were combined and FAO coordinated implementation of the joint plan.

The World Resources Institute version of TFAP claimed that all solutions to deforestation were known and that all was needed was enough money to implement them on a large scale. There were five main priority areas: forestry in land use, intended to integrate forestry and intensified agriculture; improving forest management and forest industries; meeting fuelwood needs (mainly for dry and montane areas); conserving species and ecosystems; and improving institutions, by strengthening national forestry departments, research and local involvement. But though it claimed that 'solutions outside forestry are essential' the actions recommended by the TFAP consisted almost entirely of forestry and conservation projects, and largely neglected the agriculture sector. The unstated aim was to provide a framework for the sponsoring agencies to increase financial aid to tropical forestry. There was nothing wrong in that, and indeed overseas aid to tropical forestry doubled between 1984 and 1988, but many were misled into believing it would control deforestation as well.

Early criticisms of the flaws arising from this partiality (Ross and Donovan, 1986), which also extended to the joint version, were ignored and the TFAP held centre stage for some years. It did have some positive achievements: 74 tropical countries agreed to participate in the TFAP and 62 of these invited TFAP staff to assist in national reviews of the prospects for new projects. But protests became ever more strident, criticizing the TFAP's focus on forestry projects, its traditional compartmentalized approach to land use, the lack of public participation and the poor recognition of the roles of indigenous peoples and non-governmental organizations. Finally, an external review of the TFAP was commissioned. This concluded that 'most national plans, based mainly on forestry sector reviews, simply justified increased investment in the forestry sector – a focus too narrow to adequately assess the root causes of deforestation, much less to affect them significantly.' TFAP's effect on controlling deforestation was said to be 'modest at best' and its controlling institutions had 'lost sight of these concerns as the plan has been carried out' (Winterbottom, 1990). Following these and other critical remarks moves were made to revise the TFAP, initially by changing its name to the Tropical Forestry Action Programme and broadening responsibility for coordination (Mooney, 1990; Madeley, 1990b, 1991).

Controlling tropical deforestation

The International Tropical Timber Agreement
The International Tropical Timber Agreement (ITTA) was always intended to focus on improving tropical forestry rather than controlling deforestation but nevertheless it has an important place in international affairs. Although negotiated under the auspices of the UN Conference on Trade and Development (UNCTAD) it is not a price stabilization scheme like other commodity agreements. Instead, it allows tropical hardwood producing and importing countries to work together to improve the sustainability of forest management, the effectiveness of conservation, and the diversification of the tropical hardwood trade and processing industries, with major funding for projects coming from the importing countries.

The ITTA was finalized in 1983 and ratified over the next two years. It came into force in April 1985 and in the following year it was decided to site the headquarters of its secretariat, the International Tropical Timber Organization (ITTO), in Yokohama, Japan. Since then it has implemented a number of interesting projects in producing countries (ITTO, 1990a, 1990b) and commissioned various reports, including a major study of the extent of sustainable management mentioned in Chapter 3 (Poore, 1989b). It has also been devising criteria for the important task of measuring the sustainability of forest management (ITTOc; Cassells, 1992a, 1992b). The members of ITTA (21 producing countries and 25 consuming countries) hold regular meetings, which are also notable for providing the only international public forum on tropical forests. The ITTA has received some criticism but it nevertheless provides the basis for a unique consensus approach to supporting positive developments in tropical forestry.

The Global Biodiversity Strategy
The Global Biodiversity Strategy was published jointly in 1992 by the World Resources Institute, IUCN and UNEP, essentially to expand the framework provided by the World Conservation Strategy and subsequent initiatives in the light of the new focus on biodiversity. While intended to address biodiversity conservation needs on the global scale and not just in the tropical rain forests it will, if properly implemented, greatly help to control tropical deforestation. It lists 85 guidelines for practical action at local, national and international levels by governments and non-governmental organizations while the Biodiversity Convention agreed at UNCED is being ratified and coming into force (World Resources Institute et al, 1992a, 1992b). Its scope is wider than that of the convention, so it shows how biodiversity conservation, in the humid tropics and

elsewhere, could develop in the long term. The approach taken is generally compatible with the strategy proposed in this book. To catalyse international action it calls for: an International Biodiversity Decade between 1994 and 2003; an International Panel on Bio-diversity Conservation to act as a coordinating forum and clearing house; and an Early Warning Network to monitor urgent threats to biodiversity and mobilize action against them. Otherwise, it is in the same advisory tradition as the World Conservation Strategy, urging that biodiversity conservation be integrated into national planning processes and providing lots of good ideas – perhaps too many – for improving national and international action. It will succeed or fail by how readily individual governments take up its suggestions.

Relevant achievements of the UN Conference on Environment and Development

Much was expected of the UN Conference on Environment and Development (UNCED), held in Rio de Janeiro in June 1992. As the second UN conference on the global environment it should have been a major advance over the UN Conference on the Human Envi-ronment in Stockholm in 1972, and it was in as much as the role of the developing countries was essentially neglected in Stockholm but at the forefront in Rio. The UN also succeeded in attracting 114 heads of government to Rio to attend an 'Earth Summit' in the closing days of the conference.

UNCED's two major achievements were the conventions on climate change and biodiversity that were signed there by over 150 countries. But, as mentioned in Chapters 7 and 8, the conventions were limited in scope, so judgement on them must be suspended for a few years until we see whether the world community is ready to go further and commit itself to more specific mechanisms for miti-gating global climate change and conserving biodiversity (Pearce, 1992d). If extended they could help to control deforestation, but even in its present form the biodiversity convention should lead to more conservation funding when it is implemented.

The developed countries made the mistake of trying to secure a legally binding global Forest Convention as well, but the developing countries rebuffed this in late 1991, long before UNCED, as they felt it would have compromised their sovereignty by subjecting them to unacceptable international supervision of how their forests were managed. The developed countries persisted by preparing a State-ment of Forest Principles for signature at UNCED and wanted it to contain a commitment to a future convention. But all references to this were removed in the final version of the statement, which was very general in tone. The feelings of the developing countries may be gauged from the comments of the Indian environment minister, Kamal Nath, who stated that tropical forests must remain national resources and must not be internationalized, and of Dr Mahathir Mohamad, the Prime Minister of Malaysia, who said that 'a convention would only be fair if we could also tell the North that they could not have this or that factory' (Pearce, 1992e, 1992h).

Tropical forests were mentioned in Agenda 21, the huge 800-page plan for global environmental action agreed at UNCED, but this is unlikely to be implemented due to lack of funding. The overall judgement of UNCED organizers was one of disappointment, principally at the hesitancy of the developed countries to commit themselves to modifying their own economic activities and to financing more sustainable development in developing countries – who were understandably reluctant to make any major commitments themselves (Pearce, 1992g).

Lessons from the past

Although proposed with the best of intentions previous international programmes intended to control deforestation have not had the desired effect. Chief among their drawbacks has been a partial approach. For example, the World Conservation Strategy focused mainly on conservation and the Tropical Forestry Action Plan on forestry. Agriculture has usually been neglected. Another problem with the TFAP in particular was a tendency to focus on techniques rather than treating wider underlying social, economic and policy matters as well. On the other hand, the programmes did have some good points. The World Conservation Strategy emphasized the need for tropical countries to plan their own activities within an international context and to view conservation as part of a sustainable development process. The basic principle of the International Tropical Timber Agreement was collaborative action in support of positive policies, and this is certainly an approach that can be built on in the future.

The need for a new tropical forest initiative

Given this record of failure it is surely right to question whether a new programme would do any better. However, the tropical rain forest problem is so important that this author believes we should learn from our past mistakes and make yet another attempt to tackle it.

This time, however, it is essential that the scope of action be much wider than before, taking an integrated approach that combines international support for action in the fields of agriculture, forestry and conservation, as suggested in this book. Neither international agencies nor overseas governments have the right to prescribe changes in the social and economic policies of tropical countries so this aspect will have to be be left to tropical countries themselves, but their governments should be invited to present proposals for funding by donor agencies that will improve social and economic conditions in ways that will help to control deforestation.

In contrast to the old Tropical Forestry Action Plan (TFAP) the new initiative must have a specific focus on tropical rain forests if action is to concentrate on controlling deforestation, rather than afforesting dry and montane areas, however important that is (Grainger, 1990a). A continuous global monitoring system for the tropical rain forests should also be included. This will fill a long-standing gap, give the initiative a high international profile, identify areas where action is needed and monitor the success of projects.

No new organization would be needed to implement the new initiative, the main requirement being more funding to enhance the programmes of existing agencies. Since the launch of the TFAP international responsibility for action on the different aspects of the tropical rain forest issue has become more decentralized and this is likely to happen even more in future, e.g. when the Biodiversity Convention becomes operational. It seems logical to continue this trend as long as efficient coordination can be assured, and to do this a small secretariat would be desirable. The new initiative could involve, for example, agricultural projects implemented through FAO, the development banks and national development agencies; forestry projects implemented through FAO and the ITTO; conservation projects implemented through UNESCO, the development banks and non-governmental organizations such as WWF; and the monitoring programme implemented by FAO, UNEP or a new UN agency. Other agencies and non-governmental organizations would naturally be involved as well. Coordination could be in the hands of the secretariat of the revised TFAP; alternatively that could continue as at present and a new secretariat could be set up to oversee the entire initiative, eventually embracing implementation of the relevant parts of the climate change and biodiversity conventions when they become operational.

Formulating any new initiative is likely to encounter a number of difficulties. First, the partial approach taken in previous programmes owed a lot to the inability of most international agency personnel to transcend their own disciplinary boundaries, e.g. forestry, and to the constant preoccupation of international agencies

255

themselves, and even of individual departments in those agencies, to defend their international responsibilities against competing agencies who might usurp them. The latter is a sad but omnipresent characteristic of the international environment/development scene and must be attended to urgently because of the way it limits integrated action. Second, the possibility of the TFAP being revised and continuing to play a major international role has been mentioned, but there is great distrust of the TFAP, and it might be best simply to bring it to an end and start something new. Third, and most important of all, are the attitudes of the developed and developing countries. It is regretable that some developed country governments and NGOs still favour an interventionist approach, as displayed in the abortive UNCED Forest Convention. This must be moderated if tropical countries are to agree to cooperate with a new initiative. The developed countries must appreciate the evolutionary nature of land use change and forest management and agree to support positive developments in tropical countries. If they do this then there must be a good chance that the latter will agree to cooperate.

Future trends in deforestation and logging

What is likely to happen to the tropical rain forests over the next few decades? Will most of them disappear by the early part of the next century? Are they really inexhaustible sources of timber as some have imagined? A few years ago this author built a global computer simulation model, called TROPFORM, to try to provide some answers to these questions. References have been made in earlier chapters to some of the results obtained.

Trends in deforestation

Simulations with the deforestation model in TROPFORM suggested that if modest increases in agricultural productivity are obtained, as suggested in this book, then even with continuing population growth and economic development overall deforestation rates for tropical moist forest would decline and only between 10 per cent and 20 per cent of all tropical moist forest would be lost by 2020, according to the Low and High scenarios respectively[2]. The overall deforestation rate for the 43 countries modelled fell from 4.1-6.6 million ha per annum in 1980 to 0.9-3.7 million ha per annum by 2020 in the two scenarios (Figure 9.1, Table 9.3)[3]. These countries contained 95 per cent of all tropical moist forest, and had an actual deforestation rate in 1976–80 of 5.6 million ha per annum (Lanly, 1981).

Synthesis, progress and prospects

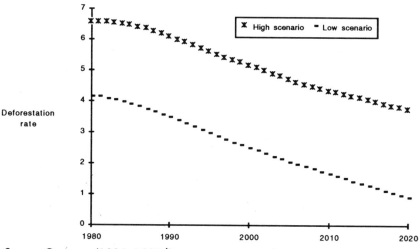

Source: Grainger (1986, 1990d)

Figure 9.1   Simulated trends in deforestation rates in the humid tropics 1980–2020 (million ha per annum)

Table 9.3.   Simulated trends in deforestation of tropical moist forest 1980–2020

| | High scenario | | | | | Low scenario | | | | |
	1980	1990	2000	2010	2020	1980	1990	2000	2010	2020
Deforestation rates (million ha per annum)										
Africa	1.552	1.486	1.207	0.885	0.866	1.036	0.935	0.670	0.559	0.427
Asia-Pacific	1.731	1.505	1.219	1.192	1.149	1.128	0.940	0.707	0.526	0.417
Latin America	3.275	3.065	2.705	2.221	1.670	1.971	1.599	1.111	0.575	0.036
Total	6.558	6.056	5.131	4.298	3.685	4.135	3.474	2.488	1.660	0.880
Forest areas (million ha)										
Africa	198.9	183.5	170.3	160.4	151.6	198.9	188.9	181.1	175.0	170.1
Asia-Pacific	239.4	222.8	209.5	197.5	185.8	239.4	228.9	220.8	214.8	210.2
Latin America	598.0	566.2	537.5	513.0	493.8	598.0	580.1	566.7	558.6	555.8
Total	1036.3	972.6	917.2	870.8	831.1	1036.3	997.9	968.7	948.5	936.1

NB. These are actual simulated values for the years specified, not averages.
Source: Grainger (1986, 1990d)

However, the deforestation model was quite simple, and deforestation rates and forest loss would be higher than this if agriculture was not sustainable, there was feedback between logging and deforestation as logging expanded in Latin America and Central Africa, governments continued to promote unnecessary deforestation for national ends, forest protection did not improve, and over-

257

seas demand for cash crops led to continuing deforestation to establish plantations (Grainger, 1986, 1990d, 1992).

The decline in overall deforestation rates masked trends in individual countries. In the High scenario, for example, deforestation rates rose initially in half of the 43 countries. The model predicted that deforestation should cease in 14 countries between 1980 and 2020 as forest cover reached critical levels and governments were forced to act to protect remaining forest, and even expand forest cover by means of new plantations. These countries were Ghana, Ivory Coast, Madagascar, Nigeria, Sierra Leone, Uganda, the Philippines, Thailand, Vietnam, Cuba, Costa Rica, Dominican Republic, El Salvador and Panama. In some of them preventative government action is already under way. Forest cover in Nigeria, Uganda and Cuba was at a critical level even in 1980, so it is hoped that their deforestation rates should be minimal in future.

Trends in logging

Trends in logging and the tropical hardwood trade were simulated by the full TROPFORM model, which included the above deforestation model as one of its components, and also took account of plantation production and regeneration after logging[4]. Various scenarios were simulated to test the effect of different trends in overseas and domestic demand for tropical hardwood. The supply side of the model included 30 leading tropical hardwood producing countries[5].

The general trend in most scenarios was for exports to peak in the first decade of the next century and then decline to below present levels by 2020 as tropical hardwood reserves were depleted by logging. So in the Base scenario, tropical hardwood exports peaked in 2005 and fell to 72 per cent of their 1980 level by 2020. In the Low scenario they peaked four years later and by 2020 were at 86 per cent of their 1980 level (Table 9.4). Overall exports in the year 2000 were 122 and 96 million cubic metres in the Base and Low scenarios respectively, compared with earlier FAO forecasts of 118 million cubic metres (Pringle, 1976) and 92 million cubic metres (FAO, 1982)[6]. The current world recession, which has led to a stagnant forest products market, could not be anticipated when the simulations were run, but one scenario showed that tropical rain forests could sustain exports at 1980 levels right through to 2020 if necessary.

Another major finding of this research was that Latin America should take over from Asia-Pacific as the top exporting region in the 1990s in the Base scenario (Figure 9.2). Latin American exports rose

Table 9.4. *Summary of trends in the tropical hardwood trade 1980–2020 simulated with the TROPFORM model: Base, Low and Constant Imports and High Deforestation scenarios (million cubic metres per annum)*

	1980	1985	1990	1995	2000	2005	2010	2015	2020
Removals									
Base	86.7	109.2	137.6	173.1	195.6	216.8	155.4	126.6	106.1
Low	86.7	104.1	126.4	155.1	173.0	181.2	176.0	144.3	113.7
Constant Imports	86.7	96.5	109.0	125.4	124.9	120.4	118.1	118.6	121.8
High Deforestation	86.7	109.2	137.6	173.2	180.7	163.8	131.4	99.4	76.6
Exports									
Base	50.6	63.4	79.2	98.4	121.9	151.0	88.8	62.6	35.9
Low	50.6	58.2	67.9	80.3	96.3	116.6	112.4	77.6	43.0
Constant Imports	50.6	50.6	50.6	50.6	50.6	50.6	50.6	50.6	50.6
High Deforestation	50.6	63.4	79.2	98.4	122.0	111.3	78.2	43.0	23.3
Imports (Base scenario)									
Japan	19.3	24.5	30.6	37.4	45.1	53.5	29.9	19.7	10.5
Europe	11.6	13.0	14.6	16.2	18.0	19.8	10.3	6.5	3.3
USA	2.7	3.1	3.4	3.9	4.3	4.8	2.5	1.6	0.8
Other Asia	10.5	13.9	18.4	24.2	31.9	41.7	25.9	19.2	11.5
Rest of the World	6.5	8.9	12.2	16.7	22.8	31.1	20.2	15.6	9.8
Exports (Base scenario)									
Africa	7.9	7.7	11.4	15.8	32.8	33.7	21.6	25.2	8.1
Asia	41.5	47.3	35.6	30.9	12.1	13.3	11.8	4.2	7.1
Latin America	1.2	8.4	32.2	51.7	77.0	104.0	55.4	33.2	20.7

NB.The above are regional/global aggregates of the mainly national trends simulated with TROPFORM. The data refer to the actual years specified and are not averages.
Scenario assumptions: Base: low economic growth, low deforestation; Low: as base but no increase in Japanese imports; Constant Imports: as Base but no increase in any imports; High Deforestation, as Base but high deforestation.

steeply in the 1990s but African exports did not follow until the early years of the next century. Indonesia remained a major Asian supply source but Malaysia declined in prominence and supplies from Burma and Papua New Guinea increased. Brazil, Colombia and Peru were the dominant Latin American suppliers by the end of the century. Zaire's exports only rose gradually. But even if overall export levels stayed constant, in all of the major producing countries *except for* Congo, Zaire, Papua New Guinea, Bolivia, Brazil, Colombia, French Guiana, Guyana, Peru and Surinam, governments would need to restrict log production before the year 2000 as tropical hardwood removals reached sustainable supply limits. Plantations would be unable to stop the decline in overall exports and removals,

Controlling tropical deforestation

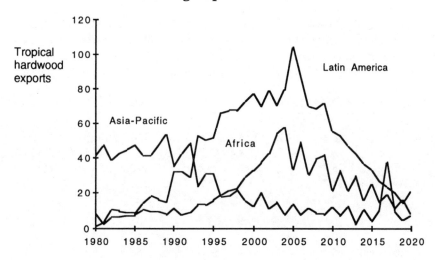

Source: Grainger (1986, 1987)

Figure 9.2 Trends in tropical hardwood exports 1980–2020 simulated with TROPFORM, Base scenario (million cubic metres per annum)

as they would not provide more than 5 per cent of all tropical hardwood production between 1980 and 2020.

Domestic consumption in the producing countries doubled in volume between 1980 and 2020 but its share of all removals declined from 42 per cent to 38 per cent. Thereafter it stayed fairly constant in volume terms because of Asian supply limitations, but its share rose as exports declined, reaching 66 per cent by 2020.

All these simulations assumed that deforestation rates followed those in the Low deforestation scenario. But if action to control deforestation was less successful and rates matched those in the High deforestation scenario instead, tropical hardwood exports were 12 per cent lower in 2010 and 35 per cent lower by 2020 than in the base scenario. By 2020, logging and deforestation had depleted commercial tropical hardwood reserves to 27 per cent of their 1980 levels in the Base scenario. The corresponding figures for the Low and Constant Imports scenarios were 31 per cent and 43 per cent, respectively (Grainger, 1986, 1987).

The main conclusions to be drawn from these modelling experiments are that a substantial area of tropical rain forest could remain by 2020 if efforts are made to increase agricultural productivity in a sustainable way, and while tropical hardwood will retain its leading role in the world forest products trade these forests are not an inexhaustible resource of commercial timber. Indeed, tropical hard-wood reserves could be seriously depleted by 2020, limiting supplies until forests regenerate. Overall exports could stay at present levels until 2020 but relatively modest rates of growth in demand could make them peak before then. Future export levels will also be greatly determined by trends in deforestation rates. The major shift to Latin America, and later Africa, as the leading exporting region has profound implications for deforestation if measures to prevent logging-deforestation feedback are not taken. Importing countries should therefore anticipate a reduced availability of tropical hard-wood after, say, 2010. The governments of producing countries should try as hard as they can to make the management of tropical rain forests more sustainable, for most tropical hardwood in the immediate future will come from these forests rather than from plantations.

The prospects for controlling deforestation

These days the future of the tropical rain forests is on the minds of a great many people, from the average person in the street to prime ministers, presidents and princes. This is a big advance from 20 years ago. International agencies, national aid organizations and non-governmental organizations are now trying their best to satisfy the widely felt need for action in the developed countries and their good intentions are to a large degree reciprocated by many dedicated but over-stretched government personnel in tropical countries. If the international programmes described above have not been all that successful, perhaps it indicates that international organizations are subject to limitations just as much as national governments. Each, it seems, can only do so much, but keeps persevering nevertheless.

Meanwhile, land use and forest cover change remorselessly in the humid tropics, driven by population growth, economic develop-ment and government policies, and forest exploitation proceeds as well to satisfy domestic and overseas demand for timber. There is no inevitable pattern of land use or forest management evolution. Chapter 4 presented two alternative scenarios, although there are others. Countries could either pass through their national land use transition relatively quickly, and retain a high forest cover, or undergo a more prolonged period of deforestation until very little forest remains. The former is more likely given sufficient investment in agriculture and a government dedicated to achieving sustainable forest management and a strong national network of protected areas. The latter will happen if farm investment is weak, land use unsustainable, forest management poor, and conservation is given a low priority.

This book has argued that there are solutions to the tropical rain forest problem, but they are not quick-acting. We can control deforestation and improve management if we start now, while there is still time. Some forest will unfortunately be lost as land use and forest management evolve, but a large area will remain nevertheless. This is not a defeatist attitude, merely a recognition that economic problems, petty international disputes, organizational shortcomings and other human limitations inevitably dampen our aspirations.

The concerns of people all over the world about the global consequences of continuing tropical deforestation seem a long way from the needs of landless peasants seeking to survive from day to day, and that is a dilemma with which we have to deal. All who are apprehensive about the future of the tropical rain forests must therefore balance global concerns with local needs. We have to recognize that these forests are the sovereign resources of tropical countries, not a global commons which is the domain of conservationists from developed countries that cleared much of their own forests in the past. Any suggestions we make as to how to tackle the problem must be framed on the understanding that people in tropical countries have the right to aspire to the same process of development that the developed nations have undergone.

If deforestation is to be controlled, the developed countries must give even stronger support to tropical countries in their efforts to improve farming, forest management and conservation. We have heard a lot about the need for a new international economic order that treats the developing countries more equitably. We now need a new international environmental order as well. For the impact of human beings on the planet has now reached such a point that if we do not begin to manage our activities on a global scale we shall soon be in trouble. Achieving effective global environmental management will demand much closer cooperation between the nations of the world than has been the case so far. The world had an opportunity to take this crucial step forward at the UN Conference on Environment and Development in Rio in 1992, but missed it. However harsh 'economic realities' may be at the moment they will be a lot harsher in future unless bold decisions are taken now. Global environmental management will demand some compromises on national sovereignty and new international financial mechanisms. Serious efforts are needed so that the next time the developed and developing nations come together a solid foundation will have been laid for agreeing on new international environmental policies.

Synthesis, progress and prospects

Most of this book has been concerned with the activities of human beings rather than the wonders of the tropical rain forests, and for good reason, since we, not they, are the problem. We are the newcomers on the scene. The tropical rain forests represent the fruits of 100 million years of evolution of the flowering plants. As recently as 40 million years ago they covered a large part of the planet, extending into high latitude regions and even as far as England[7]. While their distribution has shrunk with changing climate they still represent the ultimate in vegetation cover and have retained their vital role on the leading edge of plant evolution, a process which, according to the Cambridge botanist Paul Richards (1952), could have been fundamentally changed as a result of their destruction over the course of this century.

When we walk through these majestic forests, hear the discrete but constant chatter of the many animal inhabitants whom we rarely see, and gaze up and see the canopy towering above us, it is therefore no accident that we feel a link with both the distant past and the future, and see our own achievements in their proper perspective. We may now be in the truly awesome position of being about to change the world's climate by our polluting habits, but the fact that we can modify the natural order so quickly and on such a large scale is surely not a matter of pride.

The tropical rain forests appear rather out of place in the short-sighted world of today, where there is no room for sentiment, where vision is blinkered, and where the survival of human beings, not great ecosystems, is all important. It seems as though they have finally met their match. We can be sure that this will not always be the case. Sadly, today is what matters. The future of the tropical rain forests is literally in our hands, the hands of a species that destroys nature with the same frenetic energy as it applies to building its own machines and cities. To save these great forests requires a positive effort on our behalf, anything less will be disastrous. Somehow we must reconcile humanity's need for development and greater equity with the biosphere's need for conservation. Now we have progressed so far we must, for our own good, and ultimately our own survival, learn how to manage the planet, and this effectively means being able to manage ourselves. The tropical rain forests are one of the key tests in our learning process. For our sake, as well as theirs, we must not fail.

NOTES

Chapter 1

1. The limitations to this generalization, as shown by recent research, are explored in detail in Grainger (1993).
2. This estimate covers 59 of the 65 countries in Sommer's classification together with Puerto Rico and Vanuatu. The figures used for each country generally corresponded to Persson's estimates of 'All Closed Forest'. This was the most consistent category of data in his report although it included coniferous and bamboo forests as well as broad-leaved forests, and for some countries both dry and moist tropical forests.
3. Myers specifically estimated areas of tropical moist forest, but for only 50 of the 65 countries in Sommer's classification with the addition of Puerto Rico and he did not present any tabulated global estimate, which had to be compiled by Grainger (1984). Compensating for the countries omitted (which only account for about 1 per cent of the total area of tropical moist forest) would raise this to 983 million ha, which is very close to the Persson estimate.
4. To enable a comprehensive estimate to be made, best available data for countries not listed by FAO/UNEP were used by this author for Australia, Fiji, New Caledonia, Solomon Islands and Puerto Rico (Myers, 1980a); Reunion and Vanuatu (Persson, 1974); and Hawaii (Nelson and Wheeler, 1963).
5. However, FAO downgraded the figure of 1935 million ha (which was for 76 countries) to only 1882 million ha on the basis of better information in this second interim estimate, and gave an estimate of 1884 million ha as the area of all tropical closed forest in the new set of 87 countries in 1980.

Chapter 2

1. For a review of other classification systems for shifting cultivation see UNESCO (1983b). Other useful discussions of shifting cultivation may be found in Watters (1960, 1971); UNESCO/UNEP/FAO (1978);

Denevan (1980); Myers (1985a); Freeman (1955); Conklin (1957); Pelzer (1958); Nye and Greenland (1960); Spencer (1966); Chin (1977); Kunstadter et al (1978); Young and Wright (1980); Hatch (1982); Dove (1983); FAO (1984); and Longman and Jeník (1987).

2. Although nominally an agricultural practice, cultivation of illegal narcotic plants is discussed here since it would not fit comfortably with agricultural land uses in the previous section.

Chapter 3

1. There is considerable variation in tree and species density in the humid tropics. However the mixed dipterocarp forests of Malaysia generally contain 400–700 trees above 10cm dbh per ha, of which 200–300 will be of emergent or canopy species. Tropical rain forests in Africa and Latin America have been found to contain 350–500 trees above 10 cm dbh per ha (Whitmore, 1984: Mabberley, 1992).

2. All forest product trade statistics in this chapter and Chapter 4 come from the *FAO Forest Products Yearbook* series unless otherwise stated.

3. Roundwood equivalent volume is the estimated volume of logs required to make all the final forest products. It was calculated here by multiplying product volumes by the following conversion factors: sawnwood, 1.72; veneers and plywood, 1.81. If these factors are too low then it underestimates the proportion of all removals exported.

4. The figures in Tables 3.3 and 3.5 are necessarily approximate owing to poor data on trade flows and approximations used in calculating roundwood equivalent volume.

5. Import totals in Tables 3.3 and 3.5 differ because not all trade flows are identified in FAO statistics.

6. There is evidence from Landsat imagery that the integrity of forest reserves in north-west Ghana was maintained between 1955 and 1973 (Whitmore, 1990).

7. Protection Forest (30.3 million ha), Nature Conservation and Tourism Forest (19.0 million ha), Permanent (Timber) Production Forest (34.0 million ha), Limited Production Forest (30.0 million ha), and Conversion Forest (Production Forest that may be converted to agriculture and other purposes) (30.0 million ha) (Government of Indonesia, 1986).

8. This does not include 13.5 million ha of Indian humid tropical forests under management. As they are mainly moist deciduous forests their inclusion would give a misleading statistic for the tropical rain forests as a whole.

Chapter 4

1. A correlation coefficient (r) of 0.56 was found between the logarithm of annual deforestation rates (1976–80) and logarithm of annual population increment (1970–80) for a set of 41 countries in the humid tropics. This is not a perfect relationship but reasonable given the inaccuracy

attached to both variables. Of all the pairs of variables tested this was also the best correlation found (Grainger, 1986, 1992).
2. Now divided into Haiti and the Dominican Republic.
3. The ratio between areas under shifting and permanent cultivation in Peninsular Malaysia was 0.22 in 1974 compared with 14 in East Kalimantan in 1980.
4. This figure is based on this author's interpretation of data in Kummer (1992). Kummer uses a slightly lower figure, based on his own interpretation.

Chapter 5

1. The 13 countries were Bangladesh, Colombia, Costa Rica, Ghana, Ivory Coast, Laos, Madagascar, Malaysia, Papua New Guinea, the Philippines, Thailand, Venezuela, and North Vietnam (as it was then).
2. The 13 countries were Brazil, Burma, Colombia, Indonesia, Laos, Liberia, Madagascar, Nicaragua, Papua New Guinea, Peru, the Philippines, Thailand and Zaire.
3. FAO/UNEP included a projection of 7.5 million ha per annum for the mean deforestation rate for all tropical closed forest in 1981–85 (Lanly, 1981).
4. FAO/UNEP did not include a deforestation rate estimate for the final country in Myers' report.
5. Climax forest cover is that which is theoretically assumed to have existed prior to the emergence of human beings. It is based mainly on climate data and corresponds to maps of world vegetation types such as Figure 6.1. But we have no way of knowing whether the distribution of tropical moist forest was actually like this.

Chapter 6

1. 1 gigatonne = 10^9 tonnes, or a thousand million tonnes.
2. This refers to all tropical deforestation. The corresponding figures for deforestation in the tropical moist forests alone are 0.1–0.4 gt C per annum in 2020 and 0.4–0.7 gt C per annum in 1980. The pairs of values correspond to the low and high deforestation scenarios described in Chapter 9.

Chapter 9

1. Now also known as the World Conservation Union.
2. The model related the area of farmland in each of 43 countries in the humid tropics to changes in population, per capita food consumption and yield per hectare. The deforestation rate for each country was calculated from the extra area of farmland needed each year. Population was assumed in the model to grow in an 'S-shaped' or logistic fashion, rather than exponentially, and at an initial rate equal to the

national average in 1970–80. Two alternative scenarios, 'High' and 'Low', were tested, based on alternative assumptions about the rates of increase in food consumption per capita and yield per hectare, estimated on the basis of average regional values for 1970–80. Yield per ha grew at 1 per cent per annum for African countries (both scenarios); at 2.0 per cent per annum in the Low Scenario and 1.5 per cent in the High Scenario in Asia; and at 1.5 per cent per annum in the High Scenario and 2.0 per cent in the Low Scenario in Latin America.

3. These figures are the sum of individual annual simulations for each of the 43 countries.

4. TROPFORM projected trends in tropical hardwood consumption in Japan, Europe, USA, Other Asian Processing Countries plus Thailand, Rest of the World (People's Republic of China, Persian Gulf countries and Australia), and the African, Asian and Latin American producing countries. The consumption model and the rates of economic and population growth used for these projections were similar to those used in the IIASA Forest Sector Project's Global Trade Model (Kallio et al, 1987) which was built at the same time as TROPFORM. Demand for tropical hardwood was satisfied in an economically optimum way using a spatial allocation sub-model(based on a linear programming algorithm) to minimize the total cost of delivery (i.e. logging costs plus transport costs) to each consuming country (Grainger, 1986, 1987).

5. Since only the leading 30 producing countries were included the aggregate figures for exports may not be comparable with those in Chapter 3. These countries were Cameroon, Central African Republic, Congo, Equatorial Guinea, Gabon, Ghana, Ivory Coast, Guinea-Bissau, Ivory Coast, Liberia, Uganda, Zaire, Brunei, Burma, Indonesia, Kampuchea, Laos, Malaysia, Papua New Guinea, the Philippines, Vietnam, Bolivia, Brazil, Colombia, Ecuador, French Guiana, Guyana, Peru, Surinam and Venezuela.

6. Since TROPFORM included a larger number of consuming countries than FAO it might also be expected to predict a higher total level of consumption.

7. See Cox, Healey and Moore (1976), Chandler (1961) and Upchurch and White (1987).

BIBLIOGRAPHY

Allen, J.C. and Barnes, D.F., 1986, 'The causes of deforestation in developing countries.' *Annals of the Association of American Geographers* **75**, pp 163-84.

Allen, P.E.T., 1984, 'A quick new appraisal of the forest cover of Burma using Landsat satellite imagery at 1: 1,000,000 scale.' Rangoon: UN Food and Agriculture Organization.

Alvim, R. and Nair, P.K.R., 1986, 'Combination of cacao with other plantation crops: an agroforestry system in southeast Bahia, Brazil', *Agroforestry Systems* **4**, pp 3-15.

Anderson, D. and Grove, R., 1987, 'Introduction', in Anderson, D. and Grove, R. (eds.), *Conservation in Africa: People, Policies and Practice*, Cambridge University Press, pp 1-12.

Anon, 1971, 'Johore Tengah and Tanjong Penggerang Regional Master Plan', Hunting Technical Services Ltd.

Anon, 1987, 'Jungle is a stiff test for all adventurers', World Agriculture, Special Reprint, Omaha, Nebraska: *Omaha World Herald*, p ix

Anon, 1988a, 'Traders call for levy on tropical timber sales', *New Scientist*, 19 November.

Anon, 1988b, 'Brazil's strongarm tactics fan international opposition to dams', *New Scientist*, 19 November.

Anon, 1992, 'INBIO sets precedent in biodiversity prospecting', *Biodiversity Strategy Update*, 6. Washington, DC: World Resources Institute, Summer 1992.

Arndt, H.W., 1983, 'Transmigration: achievements, problems, prospects', *Bulletin of Indonesian Economic Studies* **19**, pp 50-73.

Asabere, P.K., 1987, 'Attempts at sustained yield management in the tropical high forests of Ghana', in Mergen, F. and Vincent J.R. (eds.), *Natural Management of Tropical Moist Forests*, New Haven: School of Forestry, Yale University, pp 47-70.

Ashton, P.S., 1978, 'The biological and ecological basis for the utilization of dipterocarps', in *Proceedings of the Eighth World Forestry Congress*, Jakarta, Indonesia.

Balick, M.J., 1985, 'Useful plants of Amazonia: a resource of global importance' in Prance, G.T. and Lovejoy, T.E. (eds.), *Key Environments–Amazonia*, Oxford: Pergamon Press, pp 339-68.

Barbier, E., Pearce, D. and Myers, N., 1990, 'Sustainable management of Amazonia' in Pearce, D.W. et al (eds.), op cit; pp 190-209.

Barbier, E., 1991, 'Tropical deforestation', in Pearce, D.W., 1991a op cit; pp 138-166.

Barham, J., 1989a, 'Law of the jungle in Brazil's tin klondike', *Financial Times*, 14 September.

Barham, J., 1989b, 'Brazilian gold output rises 23 per cent', *Financial Times*, 26 May.

Barham, J., 1989c. 'Brazilian finds weigh on tin price', *Financial Times*, 22 December.

Barham, J., 1990a, 'Tin producers study plan for further reductions in exports', *Financial Times*, 12 January.

Barham, J., 1990b, 'Brazilian court orders second tin mine closure', *Financial Times*, 18 January.

Barham, J., 1990c, 'Brazilian government aims to tame wildcat tin mine', *Financial Times*, 27 April.

Barham, J., 1990d, 'Brazilian mine road closure overruled', *Financial Times*, 23 January.

Barham, J, and Caulfield, C., 1984, 'The problems that plague a Brazilian dam', *New Scientist* 11, 10.

Barnett, A., 1989, 'Shrimps no longer small-fry in Ecuador', *Financial Times*, 18 August.

Barrett, S., 1991, 'Global warming: economics of a carbon tax', in Pearce, D.W., 1991a op cit, pp 31-52.

Barrow, C., 1988, 'The impact of hydroelectric development on the Amazonian environment: with particular reference to the Tucuruí Project' *Journal of Biogeography* 15, pp 67-78.

Barry, R.G, and Chorley, R.J., 1987, *Atmosphere, Weather and Climate*, London: Methuen.

Bethel, J. et al, 1982, 'The role of US multinational corporations in commercial forestry operations in the tropics'. A report submitted to the US Department of State, March 1982. Seattle: College of Forest Resources, University of Washington.

Blackburn, P., 1987, 'Guinea counts cost of bauxite deal', *Financial Times*, 1 April.

Blaikie, P. and Brookfield, H., 1987, 'Approaches to the study of land degradation', in Blaikie, P. and Brookfield, H, (eds.), *Land Degradation and Society*, London: Methuen, pp 27-48.

Blanford, H.R., 1958, 'Highlights of one hundred years of forestry in Burma', *Empire Forestry Review* 37, pp 33-42.

Bodegrom, A.J van and Graaf, N.R. de, 1992, 'The Celos Management System – treading softly in the forest', *Tropical Forest Management Update* 2(3), pp 5-7.

Bibliography

Bolderson, C., 1990a, 'Indonesian tin output reduction', *Financial Times*, 27 June.

Bolderson, C., 1991, 'Indonesian tin company looks to offshore wealth', *Financial Times*, 18 January.

Boonkird, S.A., Fernandes, E.C.M. and Nair, P.K.R., 1984, 'Forest villages, an agroforestry approach to rehabilitating forest land degraded by shifting cultivation in Thailand', *Agroforestry Systems* 2, pp 87-102.

Booth, H.E., 1972, 'Secondary species development', *Proceedings of the Seventh World Forestry Congress*, Buenos Aires.

Booth, H.E., 1978, 'Integrated utilization of tropical forests', *Proceedings of the Eighth World Forestry Congress*, Jakarta.

Booth, W., 1987, 'Combing the earth for cures to cancer, AIDS', *Science* 237, pp 969-70.

Boserup, E., 1965, *The Conditions of Agricultural Growth: The Economics of Agrarian Change Under Population Pressure*, Chicago: Aldine.

Boserup, E., 1981, *Population and Technology*, Oxford: Basil Blackwell.

Browder, J.O., 1985, 'Subsidies, deforestation and the forest sector in the Brazilian Amazon', World Resources Institute. Ms.

Browder, J.O., 1988, 'Public policy and deforestation in the Brazilian Amazon', in Repetto R, and Gillis, M, (eds.), *Public Policies and the Misuse of Forest Resources*; Cambridge University Press, pp 247-97.

Browder, J.O., 1989, 'Forest-based economic development in the Brazilian Amazon: A case study of frontier urban industrialization', *Journal of World Forest Resource Management* 4, pp 1-19.

Brown, K., 1991, 'Papua New Guinea peace will bring foreign investors', *Financial Times*, 20 January.

Brown, S., Gillespie, A.J.R. and Lugo, A.E., 1989, 'Biomass estimation methods for tropical forests with applications to forest inventory data', *Forest Science* 35, pp 881-902.

Brunig, E.F., von Busch, M., Heuveldop, J. and Panzer, K.E., 1975, 'Stratification of the tropical moist forest for land use planning', *Plant Research and Development* 2, pp 21-44.

Brunig, E.F., 1978, 'Tropical forests and the biosphere', in UNESCO/UNEP/FAO, *Tropical Forest Ecosystems, A State of Knowledge Report*, Paris: UNESCO, pp 33-60.

Bunyard, P., 1987, 'Dam building in the tropics: some environmental and social consequences', in Dickinson, R. (ed.), *The Geophysiology of Amazonia*; Chichester: John Wiley, pp 63-8.

Bunyard, P., 1989, 'Guardians of the Amazon', *New Scientist*, 16 December; pp 38-41.

Burgess, P.F., 1970, 'An approach towards a silvicultural system for the hill forest of the Malay Peninsula', *Malaysian Forester* 26, pp 126-34.

Burgess, P.F., 1989, 'Asia', in Poore, M.E.D. (ed.), *No Timber Without Trees*, London: Earthscan, pp 117-53.

Byrne, L., 1990a, 'Father and son jailed for murder of Mendes', *The Times*, 17 December.

Byrne, L., 1990b, 'Brazil offer to save forests in return for debts deal', *The Times*, 7 June.

Cacanindin, D.C., Micosa, L.S. Benzon, J.P. and Asilo, E.S., 1976, 'Prediction function for the estimate of clearcut areas in selectively logged-over dipterocarp forests', *Sylvatrop, Philippines Forestry Research Journal* **1**, pp 297-302.

Caldwell, J., 1976, 'Towards a restatement of demographic transition theory', *Population and Development Review* **2**, pp 321-66.

Carvajal, P.F.S., 1978, 'Heterogeneous flora from the upper Amazon and its extractive problems and experiences, in *Proceedings of the Eighth World Forestry Congress*, Jakarta.

Cassells, D., 1992a, 'ITTO develops criteria for measuring sustainable forest management', *Tropical Forest Management Update* **2**(1), p2.

Cassells, D., 1992b, 'Developments at the Yaoundé session of ITTO to promote sustainable forest management', *Tropical Forest Management Update* **2**(3), p3.

Caufield, C., 1984, *In the Rainforest*, New York: Alfred Knopf.

CDIAC, 1989, *CDIAC Communications*, Winter 1989, pp 1-3. Carbon Dioxide Information Analysis Center, Oak Ridge National Laboratory, Oak Ridge, Tennessee.

CEPA-RO, 1980, *Perspectiva Anual de Produção e Abastecimento do Território Federal de Rondônia 1979/1980*. Governo do Território Federal de Rondônia, Commissão de Planejamento Agrícola de Rondônia, Porto Velho, Brazil.

Chandler, M.E.J., 1961, *The Lower Tertiary floras of Southern England: 1, Palaeocene floras; London clay flora (Supplement)*, London.

Charney, J., 1974, Symons Memorial Lecture, London: Royal Meteorological Society.

Charney, J., 1975, 'Dynamics of deserts and drought in the Sahel', *Journal of the Royal Meteorological Society* **101**, pp 193-202.

Charters, A., 1985, 'A policy of land reform', Brazil Supplement, *Financial Times*, 20 December.

Charters, A., 1987, 'Amazon oil and gas discovered', *Financial Times*, 8 April.

Chin, S.C., 1977, 'Shifting cultivation – a need for greater understanding', *Sarawak Museum Journal* **25**, pp 107-28.

Cole, M.M., 1986, *The Savannas: Biogeography and Geobotany*, London: Academic Press.

Collins, N.M, (ed.), 1990, *The Last Rain Forests*, London: Mitchell Beazley, in association with IUCN.

Conklin, H, C., 1957, *Haunoo Agriculture, A Report on an Integral System of Shifting Cultivation in the Philippines*, Rome: FAO Forestry Development Paper No, 12.

Coone, T., 1989, '"Eco-tourism" takes off for Costa Rica', *Financial Times*, 8 November.

Cox, C.B., Healey, I.N. and Moore, P.D., 1976, *Biogeography, an Ecological and Evolutionary Approach*, (2nd Edition), Oxford: Blackwell Scientific Publications.

Bibliography

Da Cunha, R.P., 1989, 'Deforestation estimates through remote sensing: the state of the art in the Legal Amazonia', in *Proceedings of a Conference on Amazonia: Facts, Problems and Solutions*, 31 July – 2 August 1989, University of São Paulo, Brazil, pp 205-38.

Daly, H., 1991, 'Elements of environmental macroeconomics', in Constanza, R, (ed.), *Ecological Economics*, New York: Columbia University Press, pp 32-46.

Dawkins, H.C., 1961, 'New methods of improving stand composition in tropical forests', in *Proceedings, Fifth World Forestry Congress*, Seattle, 29 August – 10 September, pp 441-6.

Dawnay, I., 1989, 'Indian warriors conduct a ritual public relations exercise', *Financial Times*, 22 February.

Denevan, W.M., 1980, 'Latin America', in Klee, G.A. (ed.), *World Systems of Traditional Resource Management*, New York: Halsted Press, pp 216-7.

Detwiler, R.P. and Hall, C.A.S., 1988, 'Tropical forests and the global carbon cycle', *Science* **239**, pp 42-7.

Dickinson, R.E. and Henderson-Sellers, A., 1988, 'Modelling tropical deforestation: a study of GCM land-surface parameterizations', *Quarterly Journal of the Royal Meteorological Society* **114**, pp 439-62.

Dixon, J.A, and Sherman, P.B., 1990, *Economics of Protected Areas*, London: Earthscan.

Douglas, I., 1967, 'Natural and man-made erosion in the humid tropics of Australia, Malaysia and Singapore', Publication No, 75: 17-30, International Association of Hydrological Sciences, Washington, DC.

Douglas, I. and Spencer, T., 1985, 'Present-day processes as a key to the effects of environmental change', in Douglas, I. and Spencer, T. (eds.), *Environmental Change and Tropical Geomorphology*, London: George Allen and Unwin, pp 39-73.

Dove, M.R., 1983, 'Theories of swidden agriculture, and the political economy of ignorance', *Agroforestry Systems* **1**, pp 85-99.

Duke, J.A., 1981, *The Gene Revolution*, Office of Technology Assessment, US Congress, Washington, DC.

Dullforce, W., 1990, 'Recovery in world meat market expected to continue, says GATT', *Financial Times*, 22 February.

Dunne, N., 1989a, 'Ecuador in $9 million debt-for-nature swap', *Financial Times*, 10 April.

Dunne, N., 1989b, 'Costa Rica claims backing on debt-for-nature swaps', *Financial Times*, 8 May.

Eckholm, E., 1979, 'Planting for the Future: Forestry for Human Needs', *Worldwatch Paper* **26**. Washington, DC: Worldwatch Institute.

ECU, 1991, *Potential Socio-Economic Effects of Climate Change*, Environmental Change Unit, University of Oxford.

Eden, M.J., 1990, *Ecology and Land Management in Amazonia*, London: Belhaven Press.

Ellis, W., 1988, 'Gold fever', *Sunday Telegraph*.

EPA, 1990, *Proceedings of the Intergovernmental Panel on Climate Change, Conference on Tropical Forestry Response Options to Global Climate Change,*

9–12 January 1990, pp 93-104, Report No. 20P-2003, Office of Policy Analysis, US Environmental Protection Agency, Washington, DC.

Erfurth, T., and Rusche, H., 1976, *The Marketing of Tropical Wood, Wood Species from African Tropical Moist Forests*, Rome: FAO.

Erwin, T.L., 1983, 'Beetles and other insects of tropical forest canopies at Manaus, Brazil, sampled by insecticidal fogging', in Sutton, S.L. et al (1983) (eds.), op cit, pp 59-75.

Evans, J., 1982, *Plantation Forestry in the Tropics*, Oxford: Clarendon Press.

Ewins, P.J. and Bazely, D.R., 1989, 'Jungle law in Thailand's forests', *New Scientist*, 18 November, pp 42-6.

Eyre, S.R., 1968, *Vegetation and Soils: A World Picture*, London: Edward Arnold.

Falesi, I.C., 1974, 'O solo na Amazonia e sua relação com a definição de sistemas de produção agricola', in EMBRAPA, 1974, Reunião do Grupo Interdisciplinar de Trabalho sobre Directrizes de Pesquisa Agricola para a Amazonia Tropico Umido, Brasilia, 6–10 May.

FAO, 1958, *Tropical Silviculture*, (3 vols), Rome: UN Food and Agriculture Organization.

FAO, 1975, 'Formulation of a Tropical Forest Cover Monitoring Project', Rome: UN Food and Agriculture Organization/UN Environment Programme.

FAO, 1982, *World Forest Products: Demand and Supply 1990 and 2000, Report of an Industry Working Party*, FAO Forestry Paper No. 29, Rome: UN Food and Agriculture Organization.

FAO, 1984, *Changes in Shifting Cultivation in Africa*, FAO Forestry Paper No. 50, Rome: UN Food and Agriculture Organization.

FAO, 1985, *Tropical Forestry Action Plan*, Committee on Forest Development in the Tropics, Rome: UN Food and Agriculture Organization.

FAO, 1989, *Production Yearbook* 1987, Rome: UN Food and Agriculture Organization.

FAO, 1990, 'Interim Report on Forest Resources Assessment 1990 Project', Item 7 of the Provisional Agenda, Committee on Forestry, Tenth Session, 24–28 September 1990, Rome: UN Food and Agriculture Organization.

FAO, 1991, 'Second Interim Report on the State of Tropical Forests', presented to the Tenth World Forestry Congress, Paris, September, 1991, Rome: Forest Resources Assessment 1990 Project, FAO.

Fearnside, P.M., 1984, 'Brazil's Amazon settlement schemes', *Habitat International* 8, pp 45-61.

Fearnside, P.M., 1985a, 'A stochastic model for estimating human carrying capacity in Brazil's TransAmazon Highway colonization area', *Human Ecology* 13, pp 331-69.

Fearnside, P.M., 1985b, 'Agriculture in Amazonia', in Prance, G.T. and Lovejoy, T.E. (eds.), *Key Environments – Amazonia*, Oxford: Pergamon Press, pp 393-418.

Fearnside, P.M., 1987, 'Distribuicao de solos pobres na colonizacao de Rondônia', *Ciencia Hoje* 6, pp 74-8.

Bibliography

Fearnside, P.M., 1988, 'Jari at age 19: lessons for Brazil's silvicultural pans at Carajás', *Interciencia* **13**(1), pp 12-24.

Fearnside, P.M., 1989a, 'Deforestation and development in Brazilian Amazonia', *Interciencia* **14**, pp 291-7.

Fearnside, P.M., 1989b, 'Extractive reserves in Brazilian Amazonia', *Bio-Science* **39**, pp 387-93.

Fearnside, P.M., Tardin, A.T. and Filho, L.G.M., 1990, *Deforestation Rate in Brazilian Amazonia*, National Secretariat of Science and Technology.

Fieldhouse, D.K., 1982, *The Colonial Empires: A Comparative Survey from the Eighteenth Century*, London: Macmillan Education.

Fitzgerald, B., 1986, 'An analysis of Indonesian trade policies: countertrade, downstream processing, import restrictions and the deletion program', *CPD Discussion Paper* No. 1986-22, World Bank, Washington, DC.

Fisher, A.C., 1981, *Resource and Environmental Economics*, Cambridge University Press.

Forestry Commission, 1977, *The Wood Production Outlook in Britain – A Review*, Edinburgh: Forestry Commission.

Freeman, 1955, *Iban Agriculture: A Report on the Shifting Cultivation of Dry Rice by the Iban of Sarawak*, Colonial Research Study No. 18, London: HMSO.

Gentry, A.H., 1982, 'Neotropical floristic diversity: phytogeographical connections between Central and South America, Pleistocene climatic fluctuations, or an accident of the Andean orogeny?' *Annals of the Missouri Botanical Garden* **69**, pp 557-93.

Gentry, A.H., 1986, 'An overview of neotropical phytogeographic patterns with an emphasis on Amazonia', *Proceedings of First Symposium on the Humid Tropics*, Belém, Brazil, 12–17 November 1984 2, pp 19-35.

Gentry, A.H. and Lopez-Parodi, J., 1980, 'Deforestation and increasing flooding of the Upper Amazon', *Science* **210**, pp 1354-6.

Getahun, A., Wilson, G.F. and Kang, B.T., 1982, 'The role of trees in farming systems in the humid tropics', in MacDonald, L.H. (ed.), *Agro-forestry in the African Humid Tropics*, Tokyo: United Nations University, pp 28-35.

Gillis, M., 1988, 'Indonesia: public policies, resource management, and the tropical forest', in Repetto, R. and Gillis, M. (eds.), *Public Policies and the Misuse of Forest Resources*, Cambridge University Press, pp 43-113.

Gilpin, M.E. and Soulé, M.E., 1986, 'Minimum viable populations: processes of species extinction', in Soulé, M.E. (ed.), *Conservation Biology*, Sunderland, Massachusetts: Sinauer Associates, pp 19-34.

Goldsmith, E., 1980, 'World Ecological Areas Programme: A proposal to save the world's tropical rain forests', *Ecologist* **10**, pp 2-4.

Goldsmith, E., Hildyard, N. and Bunyard, P., 1987, 'Tropical forests: a plan for action', *Ecologist* **17**, pp 129-33.

Gollin, 1987, 'Asians harvest hope in fields of miracle rice', World Agriculture, Special Reprint, *Omaha World Herald*, pp xi, xii.

Goodland, R.J.A., 1981, 'Indonesia's environmental progress in economic development', in Sutlive, V.H. et al (eds.), *Deforestation in the Third World*, Williamsburg, Virginia: College of William and Mary.

Goodman, D., 1987, 'The demography of chance extinction', in Soulé, M. (ed.), *Viable Populations for Conservation*, Cambridge University Press, pp 11-34.

Gordon, H., 1990, 'Brazil's Indiana Jones takes on the gold hunters', *Sunday Telegraph*, 1 April.

Goudie, A.S., 1989, *The Nature of Environment*, (2nd edition), Oxford: Basil Blackwell.

Gourlay, R., 1987a, 'Philippines' land reform runs into the sand', *Financial Times*, 22 January.

Gourlay, R., 1987b, 'Aquino may lose chances for land reform', *Financial Times*, 21 April.

Gourlay, R., 1987c, 'Negros faces dilemmas of land reform', *Financial Times*, 10 July.

Gourlay, R., 1987d, 'Peasants return to Manila's bridge of protest', *Financial Times*, 25 July.

Gourlay, R., 1987e, 'Tugging at the roots of an ages old problem', *Financial Times*, 27 July.

Government of Indonesia, 1986, *Forestry Indonesia 1985/86*, Jakarta: Ministry of Forestry.

Gradwohl, J. and Greenberg, R., 1988, *Saving the Tropical Forests*, London: Earthscan.

Graham, R., 1988, 'The cocaine business', *Financial Times*, November 28.

Grainger, A., 1980a, 'The state of the world's tropical forests', *Ecologist* **10**, pp 6-54.

Grainger, A., 1980b, 'The development of tree crops and agroforestry systems', *International Tree Crops Journal* **1**, pp 3-14.

Grainger, A.,1981, 'Reforesting Britain', *Ecologist* **11**, pp 56-81.

Grainger, A., 1983, 'Improving the monitoring of deforestation in the humid tropics', in Sutton, S.L., Whitmore, T.C. and Chadwick, A.C. (eds.), *Tropical Rain Forest – Ecology and Management*, Oxford: Blackwell Scientific Publications, pp 387-95.

Grainger, A., 1984, 'Quantifying changes in forest cover in the humid tropics: overcoming current limitations', *Journal of World Forest Resource Management* **1**, pp 3-63.

Grainger, A., 1986, *The Future Role of the Tropical Rain Forests in the World Forest Economy*, D.Phil, thesis, Department of Plant Sciences, University of Oxford, Oxford: Oxford Academic Publishers.

Grainger, A., 1987, 'TROPFORM: A model of future tropical hardwood supplies', *Proceedings of the CINTRAFOR Symposium on Forest Sector and Trade Models*, Centre for International Trade in Forest Products, University of Washington, Seattle, November 3–4, pp 205-212.

Grainger, A., 1988a, 'Future supplies of high grade tropical hardwoods from tropical plantations, *Journal of World Forest Resource Management* **3**, pp 15-29.

Grainger, A., 1988b, 'Estimating areas of degraded tropical lands requiring replenishment of forest cover', *International Tree Crops Journal* **5** (1/2), pp 31-61.

Bibliography

Grainger, A., 1990a, *The Threatening Desert, Controlling Desertification*, London: Earthscan.

Grainger, A.,1990b, 'Modelling future carbon emissions from deforestation in the humid tropics', in *Proceedings of the Intergovernmental Panel on Climate Change, Conference on Tropical Forestry Response Options to Global Climate Change*, São Paulo, Brazil, 9–12 January 1990, pp 105-19, Report No. 20P-2003, Office of Policy Analysis, US Environmental Protection Agency, Washington, DC.

Grainger, A.,1990c, 'Modelling the impact of alternative afforestation strategies to reduce carbon emissions', in *Proceedings of the Intergovernmental Panel on Climate Change, Conference on Tropical Forestry Response Options to Global Climate Change*, 9–12 January 1990, pp 93-104, Report No. 20P-2003, Office of Policy Analysis, US Environmental Protection Agency, Washington, DC.

Grainger, A., 1990d, 'Modelling deforestation in the humid tropics', in Palo, M. and Mery, G. (eds.), *Deforestation or Development in the Third World? Vol, III*, Bulletin No. 349, Helsinki: Finnish Forest Research Institute, pp 51-67.

Grainger, A., 1990e, 'Overcoming institutional constraints on the global monitoring of natural resources', in Lund, H.G. and Preto, G. (eds.), *Global Natural Resource Monitoring and Assessments: Preparing for the 21st Century*, Proceedings of the International Conference and Workshop, Bethesda, Maryland: American Society for Photogrammetry and Remote Sensing, pp 1408-15.

Grainger, A., 1992, 'Population as concept and parameter in the modelling of tropical land use change', in *Proceedings of the International Symposium on Population-Environment Dynamics*, University of Michigan, Ann Arbor, 1–3 October 1990 (in press).

Grainger, A., 1993, *The Tropical Rain Forests and Man*, New York: Columbia University Press (in press).

Gray, J.W., 1983, *Forest Revenue Systems in Developing Countries*, FAO Forestry Paper No, 43, Rome.

Green, G. and Sussman, R., 1990, *Science* **248**, p 212.

Griffith, V., 1991a, 'Loud clash over tin rights', *Financial Times*, 19 April.

Griffith, V., 1991b, 'Brazilian police evict miners', *Financial Times*, 9 August.

Griffith, V., 1991c, 'Brazilian wildcat mine to reopen', *Financial Times*, 19 September.

Grigg, D.B., 1977, *The Agricultural Systems of the World*, Cambridge University Press.

Haffer, J., 1969, 'Speciation in Amazonian forest birds', *Science* **165**, pp 131-7.

Haffer, J., 1987, 'Quaternary history of tropical America', in Whitmore, T.C. and Prance, G.T. (eds.), *Biogeography and Quaternary History in Tropical America*, Oxford: Clarendon Press, pp 1-18.

Hall, A.L., 1991, *Developing Amazonia*, Manchester University Press.

Hamilton, L.S. and King, P.N., 1983, *Tropical Forest Watersheds: Hydrologic and Soils Response to Major Uses or Conversions*, Boulder, Colorado: Westview Press.

Hamzah, A., 1978, Some observations on the effects of mechanical logging on regeneration, soil and hydrological conditions in East Kalimantan', in *Proceedings of Symposium on the Long-Term Effects of Logging in Southeast Asia*, BIOTROP Special Publication No, 3 BIOTROP, Bogor, Indonesia, pp 73-8.

Hanbury Tenison, R., 1979, 'The role of the Penan in the park ecosystem', paper presented to a meeting on the results of the RGS expedition to Gunung Mulu, Sarawak, London: Royal Geographical Society.

Hansom, O.P., 1981, *A Classification of the UK Market for Sawn Hardwood*, High Wycombe: Timber Research and Development Association.

Hartshorn, G.S., Simeone, R. and Tosi, J.A. Jr., 1987, 'Sustained yield management of natural forests: a synopsis of the Palcazu Development Project in the Central Selva of the Peruvian Amazon', Ms. English translation of 'Manejo para rendimiento sostenido de bosques naturales un sinopsis del Proyecto del Palcazu en la Selva Central de la Amazonia Peruana', in Figueroa Colón, J.C., Wadsworth, F.H. and Branham, S. (eds.), *Management of the Forests of Tropical America: Prospects and Technologies*, Proceedings of a Conference, San Juan, Puerto Rico, 22–27 September 1986, USDA Forest Service, Institute of Tropical Forestry, Southern Forest Experiment Station, San Juan, pp 235-43.

Hatch, T., 1982, *Shifting Cultivation in Sarawak – A Review*, Department of Agriculture, Sarawak, Soils Division, Research Branch, Technical Paper No. 8, Kuching.

Hatch, T., and Lim, C.P. (eds.), 1978, *Shifting Cultivation in Sarawak*, Report of a Workshop, Kuching, 7–8 December 1978, Kuching: Department of Agriculture, Sarawak.

Hecht, S.B., 1980, *Some Environmental Consequences of Conversion of Forests to Pastures in Eastern Amazonia*, PhD, thesis, Berkley: University of California.

Hecht, S.B., 1983, 'Cattle ranching in the eastern Amazon: environmental and social implications', in Moran, E.F. (ed.), *The Dilemma of Amazonian Development*, Boulder, Colorado: Westview Press, 155-88.

Hekstra, G.P., 1989, 'Sea-level rise: regional consequences and responses', in Rosenberg, N.J. et al (eds.), op cit, pp 53-67.

Henderson-Sellers, A. and Gornitz, V., 1984, 'Possible climatic impacts of land cover transformations, with particular emphasis on tropical deforestation', *Climatic Change* 6, pp 231-58.

Henderson-Sellers, A. and Robinson P.J., 1986, *Contemporary Climatology*, London: Longman.

Homewood, B., 1992, 'Sacked minister attacks Brazil's green agency', *New Scientist*, 13 June.

Homma, A.O., Sá, F.T., do Nascimento, C.N.B., de Moura Carvalho, L.O.D., Mello Filho, B.M., Morreira, E.D. and Teixeira, R.N.G., 1978, 'Estudo das caracteristicas e análises de alguns indicadores técnicos e econômicos da pecuária no Nordeste Paraense, Empresa Brasileira de Pesquisa Agropecuária (EMBRAPA)', Belém.

Hoon, L.S., 1989a, 'Malaysian cocoa surges into the big time', *Financial Times*, 7 July.

Hoon, L.S., 1989b, 'Malaysia in skirmishes over environment', *Financial Times*, 27 September.

Hoon, L.S., 1989c, 'Malaysian palm oil faces fresh challenges', *Financial Times*, 15 September.

Hoon, L.S., 1990a, 'Malaysia agrees to join cocoa producers' group', *Financial Times*, 20 April.

Hoon, L.S., 1990b, 'Malaysia softens cocoa stance', *Financial Times*, 13 March.

Houghton, J.T., Jenkins, G.T, and Ephraums, J.J., 1990, *Climate Change: The IPCC Scientific Assessment*, Cambridge University Press.

Houghton, R.A., Boone, R.D., Melillo, J.M., Palm, C.A., Woodwell, G.M., Myers, N., Moore, B., and Skole, D.L., 1985, 'Net flux of carbon dioxide from tropical forests in 1980', *Nature* **316**, pp 617-20.

House of Lords, 1990, 'Tropical Forests', Report by the Select Committee on the European Communities, London: HMSO.

Humphreys, J., 1982, 'Plants that bring health – or death', *New Scientist*, 25 February.

Hunt, J., 1991, 'Environment agencies start to flex their muscles', *Financial Times*, 22 October.

IIED, 1985, 'A Review of Policies Affecting the Sustainable Development of Forest Lands in Indonesia', Jakarta.

INPE, 1980, Results summarised in Tardin, A.T. et al, 1980, 'Subprojeto desmatamento convenio IBDF/CNPQ-INPE, 1979', Relatorio No. INPE-1649-RPE/103.

INPE, 1989, Results summarised in da Cunha R.P., 1989 op cit.

ITTO, 1990a, Draft Annual Report for 1989, Yokohama: International Tropical Timber Organization.

ITTO, 1990b, Draft Report of the International Tropical Timber Council, Eighth Session, Denpasar, 16–23 May 1990, Yokohama: International Tropical Timber Organization.

ITTO, 1990c, *Guidelines for the Sustainable Management of Natural Tropical Forests*, Yokohama: International Tropical Timber Organization.

ITTO, 1992, 'Draft Criteria for Sustainable Forest Management', *ITTO Policy Development Series* Publication No. 3, Yokohama: International Tropical Timber Organization.

IUCN, 1980, *World Conservation Strategy*, Switzerland: International Union for the Conservation of Nature.

IUCN, 1991, *Caring for the Earth*, Switzerland: International Union for the Conservation of Nature.

Johnson, D.V. and Nair, P.K.R., 1985, 'Perennial crop-based agroforestry systems in north-east Brazil', *Agroforestry Systems* **2**, pp 281-92.

Johnson, R. and Fidler, S., 1991, 'G7 backing for Brazil's $1.5bn debt swap', *Financial Times*, 12 July.

Jones, P.D., Wigley, T.M.L. and Wright, P.B., 1986, 'Global temperature variations between 1861 and 1984', *Nature* **322**, pp 430-4.

Jones, P.D., Wigley, T.M.L., Folland, C.K., Parker, D.E., Angell, J.K., Lebedeff, S. and Hansen, J.E., 1988, 'Evidence for global warming in the past decade', *Nature* **332**, p 790; **333**, p 122.

Jonkers, W.B.J. and Hendrison, J., 1987, 'Prospects for sustained yield management of tropical rain forest in Surinam', in *Proceedings of a Conference on the Management of the Forests of Tropical America: Prospects and Technologies*, San Juan, Puerto Rico, 22–27 September, 1986: Institute of Tropical Forestry, USDA Forest Service, San Juan, Puerto Rico, pp 157-73.

Jordan, C.F., 1985, *Nutrient Cycling in Tropical Forest Ecosystems*, Chichester: John Wiley.

Juo, A. and Lal, R., 1977, 'The effect of continuous cultivation on the chemical and physical properties of an alfisol', *Plant and Soil* **47**, pp 567-84.

Kallio, M., Dykstra, D.P. and Binkley, C.S. (eds.), 1987, *The Global Forest Sector, An Analytical Perspective*, New York: John Wiley.

Kartawinata, K., 1979, 'An overview of the environmental consequences of tree removal from the forest in Indonesia', in Boyce, S.G, (ed.), *Biological and Sociological Basis for a Rational Use of Forest Resources for Energy and Organics*, Washington, DC: US Department of Agriculture.

Keast, A. and Morton, E.S., 1980, *Migrant Birds in the Neotropics: Ecology, Behaviour, Distribution and Conservation*, Washington, DC: Smithsonian Institution Press.

Keeling, C.D., Bacastow, R.B., Bainbridge, A.E., Ekdahl, C.A., Guenther, P.R., Waterman, L.S, and Chin, J.F.S., 1976, 'Atmospheric carbon dioxide variations at Mauna Loa Observatory, Hawaii', *Tellus* **28**, pp 538-51.

Kellman, M.C., 1969, 'Some environmental components of shifting cultivation in upland Mindanao', *Tropical Geography* **28**, pp 40-56.

Kendall, S., 1990, 'Oil and human rights a volatile mix in Ecuador', *Financial Times*, 11 December.

Kendrick, K., 1989, 'Equatorial Africa', in Campbell, D.G. and Hammond, H.D. (eds), *Floristic Inventory of Tropical Countries*, New York: New York Botanical Garden, Bronx, pp 203-16.

Kenworthy, J.B., 1970, 'Water and nutrient cycling in a tropical rain forest', in Flenley, J.R. (ed.), *The Water Relations of Malesian Forests*, Miscellaneous Series II, Department of Geography, University of Hull.

King, K.F.S., 1968, 'Agri-silviculture: the taungya system', Bulletin No. 1, Department of Forestry, University of Ibadan.

King, K.F.S., 1989, 'The history of agroforestry', in Nair, P.K.R, (ed.), op cit, pp 3-11.

Kio, P.R.O. and Ekwebelan, S.A., 1987, 'Plantations versus natural forests for meeting Nigeria's wood needs', in Mergen, F. and Vincent, J.R. (eds.), *Natural Management of Tropical Moist Forests*, School of Forestry, Yale University, New Haven, Connecticut, pp 149-76.

Knight, P., 1987a, 'Brazil's low-cost tin policy', *Financial Times*, 5 August.

Knight, P., 1987b, 'Brazilian mine sets tin target', *Financial Times*, 11 December.

Bibliography

Kunstadter, P. et al (eds.), 1978, *Farmers in the Forest: Economic Development and Marginal Agriculture in Northern Thailand*, Honolulu: East-West Centre.

Lal, R., 1981, 'Deforestation of tropical rain forest and hydrologic problems', in Lal, R. and Russell, E.W. (eds), *Tropical Agricultural Hydrology*, New York: John Wiley, pp 131-140.

Lamb, C., 1990, 'Brazil builds boom towns on fallen forests', *Financial Times*, 17 August.

Lamb, C., 1991, 'Talks raise hopes for Amazonian rainforest project', *Financial Times*, 12 December.

Lamb, C., 1992, 'Brazil attacks lack of backing for Amazon', *Financial Times*, 29 May.

Lanly, J.P. (ed.)., 1981, *Tropical Forest Resources Assessment Project (GEMS): Tropical Africa, Tropical Asia, Tropical America*, (4 vols), Rome: FAO/UNEP.

Lanly, J.P., 1982, *Tropical Forest Resources*, FAO Forestry Paper No. 30, Rome: FAO.

Ledec, G. and Goodland, R., 1988, *Wildlands: Their Protection and Management in Economic Development*, Washington, DC: World Bank.

Lennertz, R. and Panzer, K.F., 1983, 'Preliminary assessment of the drought and forest fire damage in Kalimantan Timur', Bonn: DFS German Forest Inventory Service.

Leopoldo, P.R., Franken, W., Matsui, E. and Salati, E., 1982, 'Estimativa da evapotranspiração de foresta Amazônica de terra firme', *Acta Amazônica* **12**(3), pp 23-8.

Leslie, A.J., 1987, 'The economic feasibility of natural management of tropical forests', in Mergen, F. and Vincent, J.R. (eds) op cit, pp 177-98.

Lillesand, T.K. and Kiefer, R.W., 1987, *Remote Sensing and Image Interpretation*, (2nd edition), Chichester: John Wiley.

Lintner, B., 1992, 'Burmese plunder', *Far Eastern Economic Review*, 4 June.

Liyanage, M. de S., Tejwani, K.G. and Nair, P.K.R., 1984, *Agroforestry Systems* **2**, pp 215-228.

Longman, K.A. and Jeník, J., 1987, *Tropical Forest and its Environment*, (2nd edition), London: Longman.

Lovejoy, T.E., Bierregaard, R.O., Rylands, A.B., Malcolm, J.R., Quintela, C.E., Harper, L.H., Brown, K.S., Powell, A.H., Powell, G.V.N., Schubart, H.O.R. and Hays, M.B., 1986, 'Edge and other effects of isolation on Amazon forest fragments', in Soulé, M.E. (ed.).

Low, K.S. and Goh, K.C., 1972, 'Water balance studies in Selangor, West Malaysia', *Tropical Geography* **35**, pp 60-6.

Lundgren, B.O. and Raintree, J.B., 1982, 'Sustained agroforestry', in Nestel, B. (ed.), *Agricultural Research for Development: Potentials and Challenges in Asia*, The Hague: ISNAR, pp 37-49.

L'vovich, M.I. and White, G.F., 1990, 'Use and transformation of terrestrial water systems', in Turner, B.L. II et al (eds.), *The Earth as Transformed by Human Action*, Cambridge University Press, pp 234-52.

MacArthur, R.H. and Wilson, E.O., 1967, *The Theory of Island Biogeography*, New Jersey: Princeton University Press, Princeton.

McCloskey, G., 1987, 'Colombia to sell coal mine stake', *Financial Times*, 18 March.

McCloskey, G., 1990, 'Indonesia's coal industry ready for the big time', *Financial Times*, 11 May.

McElroy, M.B, and Wofsy, S.C., 1986, 'Tropical forests: interactions with the atmosphere', in Prance, G.T. (ed.), *Tropical Rain Forests and the World Atmosphere*, Boulder, Colorado: Westview Press, pp 33-60.

McNeely, J.A., Miller, K.R., Reid, W.R., Mittermeier, R.A., Werner, T.B., 1990, *Conserving the World's Biological Diversity*, Washington, DC: International Union for the Conservation of Nature/World Resources Institute/Conservational International/ World Wildlife Fund–US/World Bank.

McNeil, M., 1964, 'Lateritic soils', *Scientific American*, November.

Mabberley, D.J., 1992, *Tropical Rain Forest Ecology*, (2nd edition), Glasgow, Blackie.

Madeley, J., 1989, 'UN agency accepts criticism of "old-fashioned" forestry policy', *Financial Times*, 5 September.

Madeley, J., 1990a, 'Keeping one jump ahead of the brown plant hopper', *Financial Times*, 19 July.

Madeley, J., 1990b, 'FAO fails to tackle "root causes" of deforestation', *Financial Times*, 29 June.

Madeley, J., 1991, 'Revamp planned for tropical forestry body', *Financial Times*, 27 March.

Magrath, W.B., and Arens, P., 1987, 'The Costs of Soil Erosion on Java – A Natural Resource Accounting Approach', Washington, DC: World Resources Institute.

Mahar, D.J., 1989, 'Government Policies and Deforestation in Brazil's Amazon Region', Washington, DC: World Bank.

Marques, J., Salati, E. and Santos, J.M., 1980a, 'A divergência do campo do fluxo de vapor dágua e as chivas na Regção Amazônica', *Acta Amazônica* **10**, pp 133-40.

Marques, J., Salati, E. and Santos, J.M., 1980b, 'Cálculo de evapotranspiração real na bacia Amazônica através do método aerológico', *Acta Amazônica* **10**, pp 357-61.

Mather, A.S., 1986, *Land Use*, London: Longman.

Mather, A.S., 1990, *Global Forest Resources*, London: Belhaven Press.

Matthews, G., 1989, 'A tribe dying in the gold rush', *The Times*, 27 January.

Matthews, R., 1990, 'Cash-starved regime courts an ecological disaster', *Financial Times*, 19 May.

Mayr, E. and O'Hara, R.J., 1986, 'The biogeographic evidence supporting the Pleistocene refuge hypothesis', *Evolution* **40**, pp 55-67.

Mdoud, Ahmed, 1989, Remarks reported in Pauley, R. 'Bangladesh "can defeat floods"', *Financial Times*, 5 December.

Bibliography

Meggers, B.J., 1987, 'Early man in Amazonia', in Whitmore, T.C. and Prance, G.T. (eds.), *Biogeography and Quaternary History in Tropical America*, Oxford: Clarendon Press, pp 153-74.

Melillo, J.M., Palm, C.A., Houghton, R.A., Woodwell, G.M. and Myers, N., 1985, 'Comparison of two recent estimates of disturbance in tropical forests', *Environmental Conservation* **12**, pp 37-40.

Meyer-Homji, V.M., 1988, 'Effects of forests on precipitation in India', in Reynolds, E.R.C. and Thompson, F.B. (eds.), *Forests, Climate and Hydrology: Regional Impacts*, Tokyo: United Nations University, pp 51-77.

Michon, G., Mary, F. and Bompard, 1986, 'Multi-storeyed agroforestry garden system in West Sumatra, Indonesia', *Agroforestry Systems* **4**, pp 315-38.

Mintzer, I.R., 1987, *A Matter of Degrees: The Potential for Controlling the Greenhouse Effect*, Washington, DC: World Resources Institute.

Molofsky, J., Menges, E.S., Hall, C.A.S., Armentano T.V. and Ault, K.A., 1984, in Veziroglu, T.N, (ed.), *The Biosphere: Problems and Solutions*, Amsterdam: Elsevier, pp 181-94.

Mooney, H.A., Vitousek P.A. and Matson, P.A., 1987, 'Exchange of materials between terrestrial ecosystems and the atmosphere', *Science* **238**, pp 926-32.

Mooney, R., 1990, 'FAO tropical forestry plan under attack', *Financial Times*, 14 June.

Moorman, F.R. and Van Wambeke, A., 1978, 'The soils of the lowland rainy climates: their inherent limitations for food production and related climatic restraints', *Eleventh Congress of the International Society of Soil Science*, Edmonton, Plenary Session Papers **2**, pp 272-91.

Moran, E.F., 1985, 'An assessment of a decade of colonisation', in Hemming, J. (ed.), Vol 2, pp 91-102.

Mori, S.A., 1989, 'Eastern, extra-Amazonian Brazil', in Campbell D.G. and Hammond H.D. (eds.), *Floristic Inventory of Tropical Countries*, New York: New York Botanical Garden, pp 427-57.

Mougeot, L.J.A., 1985, 'Alternative migration targets', in Hemming, J. (ed.), Vol 2, pp 51-90.

Murray-Brown, J., 1990, 'Indonesian villages bank on reforms to the farm economy', *Financial Times*, 2 February.

Murton, B.J., 1980, 'South Asia', in Klee, G.A. (ed.), *World Systems of Traditional Resource Management*, London: Edward Arnold.

Myers, N., 1978, 'Conservation of forest animal and plant genetic resources in tropical rain forests', in Proceedings of the Eighth World Forestry Congress, Jakarta.

Myers, N., 1979, *The Sinking Ark*, Oxford: Pergamon Press.

Myers, N., 1980, *Conversion of Tropical Moist Forests*, Washington, DC: US National Research Council.

Myers, N., 1981, 'Corn acquires genetic vigour from a wild relative', *New Scientist*, 8 January.

Myers, N., 1985a, *The Primary Source*, New York: W.W. Norton.

Myers, N., 1985b, 'A look at the present extinction spasm and what it means for the evolution of species', in Hoage, R.J. (ed.), *Animal Extinctions: What Everyone Should Know*, Washington, DC: Smithsonian Institution Press.

Myers, N., 1986, 'Tropical deforestation and a mega-extinction spasm', in Soulé, M. (ed.), *Conservation Biology: The Science of Scarcity and Diversity*, Sunderland Massachusetts: Sinauer Associates.

Myers, N., 1988, 'Threatened biotas: "Hotspots" in tropical forests', *Environmentalist* 8(3), pp 1-20.

Myers, N., 1989, *Deforestation Rates in Tropical Forests and their Climatic Implications*, London: FOE (UK).

Nair, P.K.R., 1980, *Intensive Multiple Cropping with Coconuts in India*, Hamburg: Verlag Paul Parey.

Nair, P.K.R., 1989a, 'Classification of agroforestry systems', in Nair, P.K.R. (ed.), 1989b, pp 39-52.

Nair, P.K.R. (ed.), 1989b, *Agroforestry Systems in the Tropics*, Dordrecht: Kluwer Academic Publishers.

NAS, 1980, *Research Priorities in Tropical Biology*, Washington, DC: US National Academy of Sciences.

NAS, 1983, *Changing Climate*, Report of the Carbon Dioxide Assessment Committee, Washington, DC: US National Academy of Sciences.

Nature Conservancy, 1987, quoted in Peters, R. and Lovejoy, T.E., 'Terrestrial fauna', in Turner, B.L. II et al (eds.), *The Earth as Transformed by Human Action*, Cambridge University Press, pp 353-70.

Neftel, A., Moor, E., Oeschger, H. and Stauffer, B., 1985, 'Evidence from polar ice cores for the increase in atmospheric CO_2 during the past two centuries', *Nature* 315, pp 45-7.

Nelson, R.E., and Wheeler, P.R., 1963, 'Forest Resources of Hawaii 1961', Division of Forestry, Department of Land and Natural Resources, Hawaii. In cooperation with US Forest Service, Pacific Southwest Forest and Range Experimental Station, Honolulu.

Ness, G.D. and Ando, H., 1984, *The Land is Shrinking*, Baltimore: Johns Hopkins University Press.

Nicholson, D.I., 1979, 'The effects of logging and treatment on the mixed Dipterocarp forests of South-East Asia', Rome: FAO, Ms.

Nordin, C.F. and Meade, R.H., 1982, 'Deforestation and increased flooding of the Upper Amazon', *Science* 215, pp 426-7.

Notestein, F., 1945, 'Population – the long view', in Schultz, T.W. (ed.), *Food for the World*, University of Chicago Press.

Nwoboshi, L.C., 1987, 'Regeneration success of natural management, enrichment planting, and plantations of native species in West Africa', in Mergen, F. and Vincent, J.R. (eds) op cit, pp 71-91.

Nye, P.H. and Greenland, D.J., 1960, *The Soil Under Shifting Cultivation*, Commonwealth Agricultural Bureau, Farnham Royal.

Oldfield, S., 1988, *Buffer Zone Management in Tropical Moist Forests: Case Studies and Guidelines*, Cambridge/ Gland: IUCN.

Olson, S.L. and James H.F., 1984, 'The role of Polynesians in the extinction of the avifauna of the Hawaiian Islands', in Martin, P.S, and Klein, R.G, (eds.), *Quaternary Extinctions: A Prehistoric Revolution*: Tucson: University of Arizona Press, pp 768-780.

Padua, M.T.J., 1989, 'Zoning and conservation units', in *Proceedings of a Conference on Amazonia: Facts, Problems and Solutions*, 31 July–2 August 1989, University of São Paulo, Brazil, pp 416-75.

Page, D., 1989, 'Debt-for-nature swaps', *International Environmental Affairs* **1**, 275-88.

Palmer, J., 1989, 'Management of natural forest for sustainable timber production: a commentary', in Poore M.E.D. (ed.), *No Timber Without Trees*, London: Earthscan, pp 154-89.

Palo, M., Mery, G. and Salmi, J., 1987, 'Deforestation in the tropics: pilot scenarios based on quantitative analyses', in Palo, M. and Salmi, J. (eds), *Deforestation or Development in the Third World*, Vol 1, Bulletin No. 349, Helsinki: Finnish Forest Research Institute, pp 53-106.

Parry, M.L., Carter, T.R. and Konijn, N.T. (eds.), 1988, *The Impact of Climatic Variations on Agriculture*, (2 vols), Dordrecht: Kluwer Academic Publishers.

Parry, M.L. and Carter, T.R., 1989, 'Assessing the effects of climatic change on agriculture and forestry', in Fantechi, R and Ghazi, A. (eds.), *Carbon Dioxide and Other Greenhouse Gases: Climatic and Associated Impacts*, Dordrecht: Kluwer Academic Publishers, pp 256-69.

Pearce, D.W. and Nash, C.A., 1981, *The Social Appraisal of Projects*, London: Macmillan.

Pearce, D.W., Barbier, E. and Markandya, A. (eds.), 1990, *Sustainable Development*, London: Earthscan.

Pearce, D.W. and Turner, R.K., 1990, *Economics of Natural Resources and the Environment*, London: Harvester Wheatsheaf.

Pearce, D.W. (ed.), 1991a, *Blueprint 2, Greening the World Economy*, London: Earthscan.

Pearce, D.W. 1991b, 'The global commons', in Pearce, D.W., 1991a op cit, pp 11-30.

Pearce, D.W. 1991c, 'Introduction', in Pearce D.W., 1991a op cit, pp 1-10.

Pearce, F., 1992a, 'Britain clashes with US over survival of the species', *New Scientist*, 20 June.

Pearce, F., 1992b, 'Third World wins more control of aid', *New Scientist*, 9 May.

Pearce, F., 1992c, 'Don't let us drown, islanders tell Bush', *New Scientist*, 13 June.

Pearce, F., 1992d, 'How green was our summit?' *New Scientist*, 27 June.

Pearce, F., 1992e, 'Third World fends off control on forests', *New Scientist*, 20 June.

Pearce, F., 1992f, 'Japan's billions may bypass World Bank', *New Scientist*, 13 June.

Pearce, F., 1992g, 'Earth at the mercy of national interests', *New Scientist*, 20 June.

Pearce, F., 1992h, 'Malaysia is still covered in trees', *New Scientist*, 20 June.

Pelzer, K.J., 1958, 'Land utilization in the humid tropics: agriculture', *Proceedings of the Ninth Pacific Science Congress*, Bangkok, 1957, 20, pp 124-143.

Pereira, H.C., 1973, *Land Use and Water Resources in Temperate and Tropical Climates*, Cambridge University Press.

Persson, R., 1974, 'World Forest Resources', Research Notes No. 17, Stockholm: Department of Forest Survey, Royal College of Forestry.

Peters, C.M., Gentry, A.H. and Mendelsohn, R., 1989, 'Valuation of a tropical forest in Peruvian Amazonia', *Nature* **339**, pp 655-6.

Pierce, J.T., 1990, *The Food Resource*, Harlow: Longman.

Pinker, R.T., Thompson, O.F, and Eck, T.F., 1980, 'The albedo of a tropical evergreen forest', *Quarterly Journal of the Royal Meteorological Society* **106**, pp 551-8.

Plumptre, R.A., 1972, 'Problems of increasing the intensity of utilisation of tropical high forest', *Commonwealth Forestry Review* **51**, pp 213-32.

Plumptre, R.A., 1974, 'The efficiency of sawmilling in utilising forest resources', paper presented to *Tenth Commonwealth Forestry Conference*, 1974.

Polhill, R.M., 1989, 'East Africa', in Campbell, D.G. and Hammond, H.D. (eds.), *Floristic Inventory of Tropical Countries*, New York: New York Botanical Garden, pp 5-30.

Poore, D. and Sayer, J., 1987, *The Management of Tropical Moist Forest Lands*, Switzerland: International Union for the Conservation of Nature.

Poore, M.E.D., 1989a, 'Conclusions, Recommended Action, Recent Developments, Postscript', in Poore, M.E.D., 1989b: 190-238.

Poore, M.E.D. (ed.), 1989b, *No Timber Without Trees*, London: Earthscan.

Prance, G.T., 1973, 'Phytogeographic support for the theory of Pleistocene forest refuges in the Amazon Basin based on evidence from distributional patterns in Caryocaraceae, Chrysonbalanaceae, Dichapetalaceae and Lecythidaceae', *Acta Amazônica* **3**, p 28.

Prance, G.T. 1982, 'The increased importance of ethnobotany and underexploited plants in a changing Amazon', in Hemming, J. (ed.), *Change in the Amazon Basin*, Vol, 1, pp 129-36.

Priasukmana, S., 1988, 'Indonesia as a supplier of rattan products in the international market', in Johnson, J.A. and Smith, W.R. (eds.), *Forest Products Trade: Market Trends and Technical Developments*, Seattle: University of Washington Press, pp 217-22.

Pringle, S.L., 1976, 'Tropical moist forests in world demand, supply and trade', *Unasylva* **28**(112-113), pp 106-18.

Price, C., 1989, *The Theory and Practice of Forest Economics*, Oxford: Basil Blackwell.

Princen, L.H., 1979, 'New crop development for industrial oils', *Journal of the American Oil Chemists Society* **56**(9), pp 845-48.

Proctor, J., 1983, 'Mineral nutrients in tropical forests', *Progress in Physical Geography* **7**, pp 422-31.

Proctor, J. (ed.), 1989, *Mineral Nutrients in Tropical Forest and Savanna Ecosystems*, Oxford: Blackwell Scientific Publications.

Bibliography

Pryde, E.H., Princen, L.H. and Mukherjee, K.D. (eds.), 1981, *New Sources of Fats and Oils*, Champaign, Illinois: American Oil Chemists' Society.

Pullan, R.A., 1988, 'Conservation and the development of national parks in the humid tropics of Africa', *Journal of Biogeography* **15**, pp 171-83.

Queensland Department of Forestry, 1977, *Research Report* No. 1, Brisbane, pp 72-4.

Rao, Y.S., 1988, 'Flash floods in southern Thailand', *Tiger Paper* **15**(4):1-2, Bangkok: FAO.

Raven, P.H. 1976, 'Ethics and attitudes', in Simmons, J. et al (eds.), 1976, *Conservation of Threatened Plants*, New York: Plenum Press, pp 155-79.

Raven, P.H., 1987a, 'The significance of biological diversity', Missouri Botanical Garden, St, Louis, Ms.

Raven, P.H, 1987b, 'We're killing our world', keynote address at the Annual Meeting of the American Association for the Advancement of Science, Chicago, 14 February 1987.

Reid, W.V. and Miller, K.R., 1989, *Keeping Options Alive: The Scientific Basis for Conserving Biodiversity*, Washington DC: World Resources Institute.

Repetto, R. and Gillis, M. (eds.), 1988, *Public Policies and the Misuse of Forest Resources*, Cambridge University Press.

Revelle, R., 1984, 'The effects of population growth on renewable resources', in International Conference on Population, 1983, *Population, Resources, Environment and Development*, Population Studies No. 90, New York: UN Department of International Economic and Social Affairs, pp 223-40.

Richards, J.F., 1986, 'World environmental history and economic development', in Clark, W.C. and Munn, R.E. (eds.), *Sustainable Development of the Biosphere*, Cambridge University Press, pp 53-78.

Richards, P.W., 1952, *The Tropical Rain Forest*, Cambridge University Press.

Riddell, P., 1990, 'World Bank proposes fund for environment', *Financial Times*, 4 May.

Rietbergen, S., 1989, 'Africa', in Poore, M.E.D. (ed.) op cit, pp 40-73.

Roberts, N., 1989, *The Holocene, An Environmental History*, Oxford: Basil Blackwell.

Robin, G de Q., 1986, 'Changing the sea level', in *The Impact of an Increased Concentration of Carbon Dioxide on the Environment*, New York: John Wiley.

Rocha, J., 1988, 'No land reform in Brazil charter', *Guardian*, 3 September.

Rocha, J., 1990a, 'Rape of the lost world', *Sunday Correspondent*, 7 January.

Rocha, J., 1990b, 'Brazil's army calls for "war" in Amazon', *Guardian*, 30 May.

Roche, L., 1979, 'Forestry and the conservation of plants and animals in the tropics', *Forest Ecology and Management* **2**, pp 103-22.

Ross, M.S. and Donovan, D.G., 1986, 'The World Tropical Forestry Action Plan: can it save the tropical forests?' *Journal of World Forest Resource Management* **2**, pp 119-36.

Round, P.D., 1989, 'Monitoring the conservation status of species and habitats in Thailand', *Tiger Paper*, July–September, Bangkok: FAO, pp 20-7.

Ruitenbeek, H.J., 1989, 'Social Cost-Benefit Analysis of the Korup Project, Cameroon', London: WWF.

Ruslan, M. and Manan, S., 1980, 'The effect of skidding roads on soil erosion and run off in the forest concession of Pulan Laut, South Lakimantou, Indonesia', paper presented to *Seminar on Hydrology in Watershed Management*, Surakarta, Ms.

Sader, S.A. and Joyce, A.T., 1988, 'Deforestation rates and trends in Costa Rica, 1940 to 1983', *Biotropica* **20**, pp 11-19.

Salati, E., de Oliveira, A.E., Schubart, H.O.R., Novaes, F.C., Dourojeanni, M.J, and Umana, J.O., 1987, 'Changes in the Amazon over the last 300 years', in *Proceedings of the Symposium on Earth Transformed by Human Action*, Clark University, October 1987, Cambridge University Press.

Sanchez, P.A., 1976, *Properties and Management of Soils in the Tropics*, New York: John Wiley.

Sanchez, P.A., 1979, 'Soil fertility and conservation considerations for agroforestry systems in the humid tropics of Latin America', in Mongi, H.O. and Huxley, P.A (eds.), 1979, *Soils Research in Agroforestry*, Nairobi: ICRAF.

Sanchez, P.A., 1982, 'A legume-based, pasture production strategy for acid infertile soils of tropical America', in *Soil Erosion and Conservation in the Tropics*, Madison, Wisconsin: American Society of Agronomy/Soil Science Society of America, pp 97-120.

Sanchez, P.A. and Benites, J.R., 1987, 'Low-input cropping for acid soils of the humid tropics', *Science* **238**, pp 1521-7.

Sanchez, P.A., Villachica, J.H. and Bandy, D.E., 1983, Soil fertility dynamics after clearing a tropical rainforest in Peru, *Soil Science Society of America Journal* **47**, pp 1171-8.

Sattaur, O., 1990, 'Guyana's test at high tide', *New Scientist*, 31 March.

Schimper, A.F.W., 1898, *Plant Geography Upon a Physiological Basis*, English translation by W.R. Fisher, G. Groom and I.B. Balfour, 1903, Oxford University Press.

Schlesinger, M.E., 1989, 'Model projections of the climatic changes induced by increased atmospheric temperature', in Berger, A., Schneider, S. and Duplessy, J. Cl. (eds.) *Climate and Geo-Sciences*, Dordrecht: Kluwer Academic Publishers, pp 375-415.

Schneider, S.H. and Rosenberg, N.R., 1989, 'The greenhouse effect: its causes, possible impacts, and associated uncertainties', in Rosenberg, N.R, Easterling, W.E., Crosson, P.R. and Darmstader, J. (eds), 1989, *Greenhouse Warming: Abatement and Adaptation*, Washington DC: Resources for the Earth, pp 7-34.

Schwarz, A., 1991, 'Hazed and confused', *Far Eastern Economic Review*, 31 October, pp 18, 20.

Seavoy, R.E., 1973, 'The transition to continuous rice cultivation in Kalimantan', *Annals of Association of American Geographers* **63**, pp 218-25.

Sedjo, R.A., 1983, *The Comparative Economics of Plantation Forestry: A Global Assessment*, Washington, DC: Resources for the Future.

Bibliography

Sedjo, R.A. and Solomon, A.M., 1989, 'Climate and forests', in Rosenberg, N.J., Easterling, W.E., Crosson, P.R. and Darmstader, J. (eds.), 1989, *Greenhouse Warming: Abatement and Adaptation*, Washington, DC: Resources for the Future, pp 105-19.

Setyono, S., 1978, 'The effect of selective logging as applied to the tropical rain forest in the conditions of residual stands in Indonesia', in *Proceedings of the Eighth World Forestry Congress*, Jakarta, 1978.

Setyono, S., Haerum, H., Sibero, A. and Ross, M.S., 1985, *A Review of Issues Affecting the Sustainable Development of Indonesia's Forest Land*, (Vol. 2), Jakarta.

Setzer, A.W., Pereira, M.C., Pereira, A.C. Jr, and Almeida, S.A.O., 1988, 'Relatorio de Atividades do Projeto IBDF-INPE "SEQUE" ano 1987', INPE-4534-RPE/565, Sao Paulo: Instituto de Pesquisas Espaciais.

Sherwell, C., 1990a, 'Porgera costs rise', *Financial Times*, 1 February.

Sherwell, C., 1990b, 'Australian group closes Papua copper mine as unrest continues', *Financial Times*, 8 February.

Siegenthaler, U., and Oeschger, U., 1987, 'Biospheric CO_2 emissions during the past 200 years reconstructed by deconvolution of ice core data', *Tellus* **39B**, pp 140-54.

Simberloff, D., 1986, 'Are we on the verge of a mass extinction in tropical rain forests?' in Elliott D.K, (ed.), *Dynamics of Extinction*, New York: John Wiley.

Simmons, I.G., 1989, *Changing the Face of the Earth*, Oxford: Basil Blackwell.

Simpson, E.S., 1987, *The Developing World: An Introduction*, London: Longman.

Siqueira, J.D.P., 1989, 'Sustained forest management in the Amazon: need versus research', in *Proceedings of a Conference on Amazonia: Facts, Problems and Solutions*, 31 July–2 August 1989, University of São Paulo, Brazil, pp 372-413.

Smith, N.J.H., 1981, *Rainforest Corridors: The TransAmazon Colonization Scheme*, Berkeley: University of California Press.

Soekotjo, 1977, 'Research on the implementation of the Indonesian selective felling system and intensification of the management of forest exploitation in East Kalimantan', Yogyakarta: Faculty of Forestry, Gadjah Mada University.

Sommer, A., 1976, 'Attempt at an assessment of the world's tropical forests', Unasylva 28(112-113), pp 5-25.

Soulé, M.E., 1980, 'Thresholds for survival: maintaining fitness and evolutionary potential', in Soulé, M.E. and Wilcox, B.A. (eds.), *Conservation Biology: An Evolutionary-Ecological Perspective*, Sunderland, Massachusetts: Sinauer Associates, pp 151-70.

Soulé, M.E., 1987, 'Introduction', in Soulé M.E. (ed.), *Viable Populations for Conservation*, Cambridge University Press, pp 1-10

Spencer, J.E., 1966, 'Shifting Cultivation in Southeastern Asia', Publications in Geography No, 19, University of California.

Sterba, J.P., 1987, 'Malaysian tribe fights to preserve forests, win native rights', *Wall Street Journal*, 22 July.

Stoler, A., 1975, *Garden Use and Household Consumption Pattern in a Javanese Village*, PhD, thesis, New York: Department of Anthropology, Columbia University.

Suarez de Castro, F. and Rodriguez, A., 1955, 'Perdidas por erosion de elementos nutritivos bajo differentes dubiertes vegetales y con varios practicas de conservacion de suelos', Federacion Nacional de Cafeteros de Colombia, Bol.Tec No.14.

Sun, M., 1988, 'Costa Rica's campaign for conservation', *Science* **239**, pp 1366-9.

Suryatna, E.S. and McIntosh, J.C., 1980, 'Food crops production and control of *Imperata cylindrica L*, (Beauv) on small farms', in *Proceedings of the Biotropical Workshop on Alang-Alang*, Bogor: Special Publication No, 5, Biotrop, pp 135-47.

Swanson, T., 1991, 'Conserving biological diversity', in Pearce, D.W., 1991a op cit, pp 181-208.

Synnott, T., 1989, 'South America and the Caribbean', in Poore, M.E.D. (ed.) op cit, pp 74-116.

Takeuchi, K., 1981, 'Mechanical Processing of Tropical Hardwood in Developing Countries: A Review of Issues and Prospects, with Special Emphasis on the Plywood Industry in Southeast Asia', Washington, DC: World Bank, Ms.

Tang, H.T., 1987, 'Problems and strategies for regenerating dipterocarp forests in Malaysia', in Mergen, F. and Vincent, J.R. (eds.), *Natural Management of Tropical Moist Forests*, New Haven, Connecticut: School of Forestry and Environmental Studies, Yale University, pp 23-45.

Tangley, L., 1988, 'Agency acts to protect biological diversity', *Bioscience* **38**(2), p 88.

Tardin, A.T., dos Santos, A.P., Moraes Novo, E.M.I. and Toledo, F.L., 1978, 'Projetos agropecuários da Amazonia; destamento e fiscalização-relatório, A Amazõnia Brasiliera em Foico 12, pp 7-45.

Tardin, A.T. et al, 1980, Subprojeto desmatamento convênio IBDF/CNPQ-INPE, 1979, Relatorio No, INPE-1649-RPE/103.

Teitelbaum, M., 1975, 'The relevance of demographic transition theory for developing countries', *Science* **188**, pp 420-5.

Terra, G.T.A. 1954, 'Mixed garden horticulture in Java', *Malaysian Journal of Tropical Geography* **4**, pp 33-43.

Tyler, C. and Kendall, S., 1987, 'Colombians swoop on drug ring', *Financial Times*, 18 June.

U Kyaw Zan, 1953, 'Note on the effect of 70 years treatment under selection system and fire protection in the Kanyi Reserve', *Burmese Forester* **3**(1), pp 11-8.

UN, 1984, *International Tropical Timber Agreement*, 1983, New York: United Nations.

UN, 1992, 'Report of the Intergovernmental Negotiating Committee for a Framework Convention on Climate Change', 15 May 1992, Annex 1, New York: United Nations Framework Convention on Climate Change.

Bibliography

UNEP, 1992, 'Convention on Biological Diversity', Conference for the Adoption of the Agreed Text of the Convention on Biological Diversity, Nairobi, 22 May 1992, Nairobi: UN Environment Programme.

UNESCO, 1981, 'Biosphere Reserves', MAB Information System Publication No. SC-81/WS/61, Paris: UN Educational, Scientific and Cultural Organization.

UNESCO, 1983a, *Swidden Cultivation in Asia, Vol 1. Content Analysis of the Existing Literature*, Paris: UN Educational, Scientific and Cultural Organization.

UNESCO, 1983b, *Swidden Cultivation in Asia, Vol 2. Country Profiles*, Paris: UN Educational, Scientific and Cultural Organization.

UNESCO/UNEP/FAO, 1978, *Tropical Forest Ecosystems: A State of Knowledge Report*, Paris: UN Educational, Scientific and Cultural Organization.

Upchurch, G.R. and Wolfe, J.A., 1987, 'Evidence from fossil leaves and wood', in Friis, E.M., Chaloner, W.G. and Crane, P.R. (eds.), *The Origins of Angiosperms and their Biological Consequences*, Cambridge University Press, pp 75-105.

USDA, 1975, *Soil Taxonomy*, USDA Agriculture Handbook No. 436, Washington, DC: US Department of Agriculture.

Vanvolsem, W., 1990, 'Brazilian gold rush threatens bloodbath for Amazon tribe', *Sunday Telegraph*, 14 January.

Vergara, N.T. and Nair, P.K.R. 1985, 'Agroforestry in the South Pacific Region: an overview', *Agroforestry Systems* **3**, pp 363-79.

Von Droste, B., 1988, 'The role of biosphere reserves at a time of increasing globalization', in Martin, V. (ed.), 1988, *For the Conservation of Earth*: Golden, Colorado: Fulcrum, pp 89-93.

Wadsworth, F.H., 1987, 'Applicability of Asian and African silviculture systems to naturally regenerated forests of the Neotropics', in Mergen, F. and Vincent, J.R. (eds.) op cit, pp 93-111.

Ward, R.C. and Robinson, M., 1990, *Principles of Hydrology*, London: McGraw Hill.

Watters, R.F., 1960, 'The nature of shifting cultivation', *Pacific Viewpoint* **1**, pp 59-64.

Watters, R.F., 1971, *Shifting Cultivation in Latin America*, Rome: FAO.

Weaver, P.L., 1987, 'Enrichment plantings in Tropical America', in *Proceedings of a Conference on the Management of the Forests of Tropical America: Prospects and Technologies*, San Juan, Puerto Rico, 22–27 September 1986, San Juan, Puerto Rico: Institute of Tropical Forestry, USDA Forest Service, pp 259-78.

Webb, B., 1989, 'Calling on the world to help', *The Times*, 7 October.

Webb, L.J. and Higgins, H.G., 1978, 'The impact of the new ecology in the development of the tropical rain forests', in *Proceedings of the Eighth World Forestry Congress*, Jakarta, 1978.

Weidelt, H.J. and Banaag, V.S., 1982, *Aspects of Management and Silviculture of Philippine Dipterocarp Forests*, Eschborn: German Agency for Technical Cooperation.

Wells, D.R., 1971, 'Survival of the Malaysian bird fauna', *Malayan Nature Journal* **24**, pp 248-56.

Wells, D.R., 1974, 'Resident birds', in Lord Medway and Wells, D.R. (eds.), *Birds of the Malay Peninsula*, London: Witherby.

Wells, S. and Edwards, A., 1989, 'Gone with the waves', *New Scientist*, 11 November.

Westoby, J., 1987, *The Purpose of Forests*, Oxford: Basil Blackwell.

Westoby, J., 1989, *Introduction to World Forestry*, Oxford: Basil Blackwell.

Whitmore, T.C., 1984, *Tropical Rain Forests of the Far East*, (2nd edition), Oxford: Clarendon Press.

Whitmore, T.C., 1990, *An Introduction to Tropical Rain Forests*, Oxford: Clarendon Press.

Whittaker, R.H., 1975, *Communities and Ecosystems*, London: Collier Macmillan.

Whittaker, R. and Likens, G.E., 1975, in Leith, H. and Whittaker, R.H. (eds.), *Primary Productivity of the Biosphere*, New York: Springer, pp 281-302.

Wiersum, K.F., 1984, 'Surface erosion under various tropical agroforestry systems', in O'Loughlin, C.L. and Pearce, A.J. (eds.), *Symposium on Effects of Forest Land Use on Erosion and Slope Stability*: Honolulu: East-West Center, pp231-9.

Wigley, T.M.L. and Jones, P.D., 1985, 'Influences of precipitation changes and direct CO_2 effects on streamflow', *Nature* **314**, pp 149-52.

Wigley, T.M.L. and Schlesinger, M.E., 1985, 'Analytical solution for the effect of increasing CO_2 on global mean temperature', *Nature* **315**, pp 649-52.

Williams, C.N., 1975, *The Agronomy of the Major Tropical Crops*, Kuala Lumpur: Oxford University Press.

Williams, M., 1989, *Americans and their Forests: A Historical Geography*, Cambridge University Press.

Williams, M., 1990, 'Forests', in Turner, B.L. II et al (eds.), *The Earth as Transformed by Human Action*, Cambridge University Press, pp 179-201.

Winterbottom, R., 1990, 'Taking Stock: The Tropical Forestry Action Plan After Five Years', Washington, DC: World Resources Institute.

Wilson, E.O. (ed.), 1988, *Biodiversity*, Washington, DC: National Academy Press.

Wilson, E.O., 1989, 'Threats to biodiversity', *Scientific American*, September 1989.

Wilson, G.F. and Kang, B.T., 1980, 'Developing stable and productive biological cropping systems for the humid tropics', *Proceedings of the Conference on Biological Agriculture*, Wye College, London, 22–26 August.

WMO, 1986, *Report of the International Conference on the Assessment of the Role of Carbon Dioxide and Other Greenhouse Gases in Climate Variations and Associated Impacts*, Villach, Austria, 9–15 October 1985, WMO No. 661, Geneva: World Meteorological Organization.

Woodwell, G.M., Whittaker, R.H., Reiners, W.A., Likens, G.A., Delwiche, C.C. and Botkin, D.B., 1978, 'The biota and world carbon budget', *Science* **199**, pp 141-5.

Bibliography

World Bank, 1978, *Forest Sector Policy Paper*, Washington DC.

World Bank, 1980, *World Development Report*, Oxford University Press.

World Commission on Environment and Development, 1987, *Our Common Future*, New York: United Nations.

World Resources Institute, 1987, *World Resources*, Washington DC: World Resources Institute.

World Resources Institute, 1988, *World Resources* 1988-89, New York: Basic Books.

World Resources Institute, World Bank and UN Development Programme, 1985, *Tropical Forests: A Call for Action*, Washington DC: World Resources Institute.

World Resources Institute, World Conservation Union and UN Environment Programme, 1992a, *Global Biodiversity Strategy*, Washington, DC: World Resources Institute.

World Resources Institute, World Conservation Union and UN Environment Programme, 1992b, *Global Biodiversity Strategy: A Policy-maker's Guide*, Washington, DC: World Resources Institute.

Wyatt-Smith, J., 1963, Manual of Malayan Silviculture, *Malayan Forest Record* No. 23, Kuala Lumpur: Forest Department.

Wyatt-Smith, J., 1987, 'Problems and prospects for natural management of tropical moist forests', in Mergen, F. and Vincent, J.R. (eds.) op cit, pp 5-22.

Young, A. and Wright, A.C.S., 1980, 'Rest period requirements of tropical and subtropical soils under annual crops', in *Report of the second FAO/UNFPA Expert Consultation on Land Resources for the Future*, Rome: UN Food and Agriculture Organisation, pp 197-268.

Zimmerman, B.L. and Bierregard, R.O., 1986, 'Relevance of the equilibrium theory of island biogeography with an example from Amazonia' *Journal of Biogeography* **13**, pp 133-43.

Zon, R. and Sparhawk, W.N., 1923, *Forest Resources of the World*, New York: McGraw Hill.

INDEX

Index

makore, 71
Malaysia, *see also* Sabah and
  Sarawak
  forest policy, 117
  forest reserves, 88
  forests in, 40
  oil palm plantations, 62
  Peninsular, 40
  role in tropical hardwood trade,
    80
  rubber plantations, 61
  selective logging systems, 83
  tin mining, 63
  tropical hardwood exports, 78,
    80, 113
Malesia, 40
Malesian archipelago: *see* Malesia
Man and Biosphere Programme
  (UNESCO), 247
mangrove forest , 36
Mauritius, 40
medicines, threat to forest, 154–5
Meliaceae, 81
Mendes, Chico, 61, 201
meranti, 71
methane, 166
Mexico, 40
migration, 51–3, 57–9, 95
minimum critical reserve size, 206
minimum viable population, 150,
  206
mining and deforestation, 63–6
minor forest products
  economics of, 201
  possible new integrated
    harvesting methods, 201–2
  threat to, 153–5
monitoring,
  deforestation, 48
  degradation, 48
  global environmental change,
    23–4
  logging, 70
  national forest, 45
  of tropical forests, continuous,
    140–4, 238–40
  see also remote sensing
monsoon forest, *see* tropical moist
  deciduous forest

montane rain forests, 37–8
Montreal Protocol, 239
Myers, Norman, 41, 44, 124, 127–8,
  131–3, 136–9, 141, 203

narcotics, cultivation of illegal, 67
national conservation stategies,
  226–7, 250,
national land use policy, integrated,
  224–5
national land use transition, 104–9
  experience of UK and USA,
    105–7
  analysis for humid tropics,107–8
  threats to new equilibrium,
    107–8
National Oceanographic and
  Atmospheric Administration
  (NOAA), 126, 132, 140
natural gas, 65
New Guinea, 40, 56, 82, 102, 109
Nepal, 38
Netherlands, 75
Nicaragua, 40, 59, 249
Nigeria
  alley cropping in, 184
  conservation status, 249
  forest loss, 136, 140
  forests in, 40
  petroleum exports, 65
  planted tree fallows in, 180
  tropical hardwood exports, 113
nitrous oxide, 166
non-governmental organizations,
  210–11, 223, 232
nutrient cycling, 39

obeche, 71
Other Asian Processing Nations, 75,
  78, 121
oil
  essential, 153
  palm, 62
  petroleum, *see* petroleum
okoume, 71
opium, 52, 57
option values, 212
orangutan, 150
Organization of Petroleum

304